Studies in Innovation in the Steel and Chemical Industries

by
J. A. ALLEN

*Professor of Chemistry
in the University of Newcastle,
New South Wales, Australia*

Augustus M. Kelley · Publishers
New York · 1968

Published in the United States, 1968
by AUGUSTUS M. KELLEY · PUBLISHERS
24 East 22nd Street, New York, N.Y. 10010

G.B. SBN: 7190   0293   1

Printed in the Republic of Ireland
by Hely Thom Limited Dublin

CONTENTS

TABLES

FIGURES

PREFACE

Since the Second World War, a considerable body of literature on industrial invention and innovation has been published. For the most part, this has dealt with general problems of innovation with particular reference to research and development and to management, or has been concerned with case studies often limited in depth or breadth with emphasis predominantly on economic or social aspects.

The present studies on polythene, Terylene and oxygen steelmaking, which are centred geographically in the United Kingdom, are based on the premise that meaningful studies of this kind should be related in some depth to the science, technology and industrial art from which they emerged, and projected against their contemporary economic, social and political environment. Their structure reflects these dual themes in this order. Though each is fully self-contained, they are presented on a more or less common historical pattern, and some speculative comparisons between them are attempted.

In sketching the relevant scientific and technological backgrounds, no attempt has been made to give complete, professionally satisfying accounts, and experts may validly point to some omissions and deficiencies. The aim has been to provide sufficient technical detail in an effort to ensure an understanding of the generation and flowering of ideas and of their subsequent development and industrial exploitation for the reader who is neither wholly bereft of scientific or technological knowledge, nor possessed of a substantial background in these fields.

The three subjects are uncomfortably close to us in time, so that any judgments offered must necessarily be somewhat tentative. The information on which these studies rest is a mixture of documented fact and material obtained in countless interviews and discussions, in a number of cases with those who participated actively in some of the events. The correlation and integration of this diverse body of knowledge obtained from these widely differing sources and any opinions expressed as a result are, of course, wholly the author's responsibility.

Many people, far too numerous to mention individually, generously and patiently contributed their time to answer or comment on many questions in a manner which made this part of my task a stimulating and memorable experience. At the risk of appearing ungracious to many, my special thanks are due to Drs R. O. Gibson and E. Hunter and to the late Mr J. R. Whinfield who read and commented upon the drafts of Sections II and III, respectively. The lengthy quotation from the paper by J. W. Hill and W. H. Carothers is reproduced by kind permission of the American Chemical Society.

Finally, it is a pleasure to acknowledge my indebtedness to the University of Manchester for the award of a Simon Senior Research Fellowship during 1965 concurrently with a period of study leave from my own university, and to Professor B. R. Williams for his continuing wise counsel and valuable criticisms of the text as a whole.

J.A.A.

Manchester and
Newcastle, N.S.W.
June 1967

I

INTRODUCTION

Innovation and the idea of progress

One of the significant themes in the study of history is the idea of progress. This arose in the sixteenth and seventeenth centuries as a purely Western idea against the emergence of the New World and developed as the scientific revolution began to gather pace. It came to embrace not only the concept of man securing increased control over his material environment, but predicated moral and social progress as well. In operational terms, it reduces to the introduction into widespread use of new products, new processes, new ways of doing things and new forms of organization—in brief, to innovation. As a generic term, innovation may be applied to many areas of human endeavour, to agriculture, industry and medicine, and to a variety of other intellectual, cultural, social, economic and political pursuits.

The dictionary definitions of innovation and invention often overlap, so that it is essential at the outset to clarify the distinction we shall wish to draw between these two terms. An invention may be defined as the creation of an idea and its first reduction to practice, by which we mean, in the case of a machine, the construction of a single or small number of demonstration units; in the case of a process, the erection of a laboratory or semi-technical scale plant; and for a product, the preparation of trial quantities of the new material, the amount depending on its nature and intended use. One aspect of innovation is concerned with the translation of an invention into general use and, in this sense, an invention may be treated as the first stage of the more extensive and complex process of innovation. There are, however, other important aspects of innovation, for example, the introduction of new ways of organizing capital and labour and new methods of process control, materials handling, power and fuel utilization, design, marketing and distribution.

Two essential elements in an invention are novelty and usefulness, both of which may range widely in degree. The common use of the term, invention, is confined to strikingly new products or processes, but these are only the peak statistics—that part of the invention iceberg which shows above the surface. Below them, lies the vast array of smaller and less well known inventions, some patented and others not, which in total magnitude and effect may exceed that of the better known, named examples. These smaller contributions are an important component of evolutionary as distinct from revolutionary technical change.

Outstanding new products and processes vary greatly in their origins and novelty. Some arise from everyday experience cogently observed, some from deliberately contrived trials of a range of possibilities, while others stem from scientific research either by accident or from work directed to a particular end. Many new products are new only in application or in the nature of their formulation or presentation. The initial stimuli for inventions generally may come from a variety of sources including the ultimate consumer, the sales force, the process operators and supervisors, plant designers and contractors, and research and development departments or institutions which have been established to foster the flow of possibilities of this kind or to enhance the store of knowledge on which some will vitally depend. Management, too, is often a potent source from which novel ideas may flow.

The majority of products incorporating new compositions of matter, and new processes based upon discoveries in science are increasingly the direct or indirect result of scientific research. Because of his primary concern with new knowledge of the material world, the scientist belongs to the professional group which in a modern, industrial society is likely to be involved in the creation of products and processes of this type. Whether such products and processes at the discovery or invention stage are candidates for development and ultimately widespread exploitation depends not only on their intrinsic merit, but also on the nature and needs of their total environment. It is not surprising, therefore, that some major discoveries are rapidly developed and exploited, while others languish for years until the environmental needs or conditions become more favourable, or can be made so.

Some possible misconceptions

Every invention involving the creation of an idea and its reduction to practice makes a contribution to knowledge, however small. If we accept the common usage of the word, research, as that activity of extending the bounds of knowledge, invention is properly included in this wider term. But to equate invention and research without qualification, or to assume that there exists some simple proportionality between them is seriously to oversimplify the quite subtle relationships which exist between pure and applied science.

The most commonly advanced scheme of innovation is in the form of a linear sequence, beginning with research at the origin and finishing with distribution. Such a scheme might be as follows:

Research → Development → Investment → Construction → Production → Distribution.

There are any number of particular cases which fit such a scheme, especially when outstanding new products have been generated within the science-based industries, but even in these cases it does not follow that the rate of the overall process is necessarily increased solely by increasing the driving pressure at the research end, or by applying suction at the right hand end—the analogy of finding or creating a market for the product. The rate of the total operation depends ultimately on the capacity to perform, or have performed, all the necessary intermediate stages at appropriately coordinated rates and to manage successfully the progress from stage to stage.

To the extent that such a representation implies that all the stages are conducted in the one organization, even where in other respects it is a fair picture of the sequential steps, it fails to reflect the real situation. Though there are undoubtedly cases in which virtually the whole of the operation has been carried out by and in a single firm or organization, there are in total many more in which one or more of the stages has been prosecuted or obtained by external contract, purchase, royalty or distribution agreement or in other ways. Indeed, if this were not a common practice both nationally and internationally, the spread of industrial innovation would be severely inhibited. The criteria by which any particular situation or set of decisions of this kind should be judged are primarily economic ones, modified if necessary by short or long term considerations.

There are, however, many examples in which the linear scheme only very imperfectly represents, or does not describe at all, the innovation effort in a firm or an industry. In a number of so-called traditional industries, the economic advantage possibly over a sustained period of time may lie not in the rapid and frequent introduction of wholly new products or revolutionary processes, but rather in devising new ways of doing things, of organizing labour and utilizing existing capital equipment, of improving material efficiency, or of promoting new markets or designs, or, generally, by decreasing production costs by a combination of many practices of these kinds. While attention to these aspects is important in most industries whether science-based or traditional, their importance relative to the invention, development and the introduction of wholly new products and processes may be different from industry to industry.

The failure to appreciate this point has led to frequent and sometimes invidious comparisons between the higher ratio of research and development expenditure to output in the high-growth compared with the low-growth industries. In the former category, the aircraft, chemical, pharmaceutical, electronics and precision instrument industries are often compared with the iron and steel, shipbuilding, pottery and ceramics industries in the latter. In such comparisons, there is often a stated or implied assumption that there exists a simple cause and effect relationship between the research and development expenditure ratio and rate of growth, and that low-growth industries could be transformed into high-growth ones by simply increasing their research and development expenditures. This proposition is, in general, false because the potential for research and development to induce a high rate of growth varies greatly from industry to industry. This view is supported by the observation that remarkably similar research and development expenditure ratios are found for particular industries in most advanced countries of the world. In brief, the high growth industries spend large sums on research and development because they find this profitable.

The present studies

Since the Second World War, there has appeared a considerable body of literature dealing with various aspects of industrial invention and innovation. Broadly speaking, these publications fall into three categories: (i) those dealing with innovation in its general industrial or national aspects; (ii) problems of research and development often with special reference to management and to enhancing industrial creativity; and (iii) case studies ranging from comprehensive but rather superficial comparisons of inventions to more detailed examinations of particular examples. In some, the emphasis has been predominantly on economic aspects, a limited number has been more concerned with scientific and technological issues, but few have dealt in both breadth and depth with the total situation from the pre-invention stage through to the point of full scale production. Such analyses appear to be required if valid general assessments are to be made.

The task of doing this for a large number of cases is a daunting one and there are obvious advantages in trying at the outset to select examples which might prove most informative. In the light of the comments in the last section, one of the first requirements is to try to include at least two industries, one of high and the other of the low growth type, but avoiding the extremes of either type. The chemical and the iron and steel industries adequately meet this aim. Moreover, they are both of central economic importance in any industrialized nation and bear something like the same general relationship to their total national environment. Both are highly diversified, so that it is desirable to narrow much further the choice within them.

A rigid distinction between innovations involving new products and those concerned with new processes is often difficult to draw because the former will sometimes entail the latter, while a new process may be directed to the manufacture of either a new or an existing product. In these circumstances, there are advantages in trying to include cases in which a variety of these situations is involved. It is also preferable in a limited group of studies that the cases chosen should not be too divergent in type; for example, a comparative study of a narrow spectrum antibiotic and a method of producing a common metal in millions of tons for a vast range of purposes is unlikely to be particularly instructive. Similarly, it is scarcely profitable to place in apposition the design, development and construction of a new aeroplane composed of a very large number of individual components with an innovation involving a basic raw material which is supplied to other industries.

These and other considerations of a similar kind led to the choice of the three cases examined in this book, namely, polythene,[1] Terylene[2] and oxygen steelmaking, the first two essentially within the framework of the chemical industry and the third from the iron and steel industry. Commercially, all three belong to and effectively span the period since the beginning of the Second World War; all are concerned with basic materials in the high tonnage range at about the same point in their respective manufacturing sequences; and they include a useful mixture of two new and one old product and at least two novel technologies.

[1] Also called polyethylene in the U.S.A. and some European countries.
[2] Dacron in the U.S.A.

Geographically, these studies are centred on the United Kingdom and are related, where appropriate, to similar developments elsewhere in the world. This is largely a matter of convenience, but it also permits some comparisons to be made between innovations derived from both local inventions and those imported in developed form from abroad. Each of the studies is fully self-contained, but they are presented on a more or less common historical pattern. A good deal of attention is given to the relevant aspects of the environment in which the particular inventions were made and subsequently developed and, where the exploitation has depended on cognate developments in other fields, some discussion of these is also included. In this way, it has been possible to provide deliberately some internal standards for comparison and to increase effectively the number of circumscribed cases beyond the nominal figure.

Studies in industrial innovation, if they are to be meaningful, necessarily have to be related in some detail to the science, technology and industrial art of their time and projected against the economic, social and political conditions which prevailed. The structure of these studies largely reflects these dual themes in this order. In sketching the scientific and technological background, no attempt has been made to give complete, professionally satisfying accounts and experts in these fields may properly point to some deficiencies. On the other hand, there can be no escape from the need to provide a connected discussion in sufficient technical detail to ensure an understanding of the generation and flowering of ideas and of their subsequent development and exploitation. In so far as this can be done, an attempt is made to tread a middle path between the needs of a reader who has little or no scientific or technological background and one who has a good deal of knowledge and experience in these fields.

II
POLYTHENE

Introduction

Chemistry as a science is concerned with two broad areas of endeavour: one is the elucidation and correlation of the properties of chemical substances, and the other is the description and prediction of the dynamics of chemical change. Modern chemical industry, which rests heavily on its cognate science, has as its central technical aim the manufacture of an ever widening array of materials of new chemical constitution from a limited number of starting materials. In this respect, the industry differs from many other sectors of manufacturing in which the task is primarily one of physically reshaping a restricted number of materials for an increasing range of particular uses. One achievement of the chemical industry in modern times which has had a major impact on everyday life has been the introduction of a variety of polymeric substances from which the ubiquitous plastics are compounded.

Polymers may be grouped in three broad classes:

(i) natural polymers, e.g. silk, wool, cotton, natural rubber;

(ii) derived polymers obtained by chemical modification of natural polymers, e.g. nitrocellulose, cellulose acetate;

(iii) synthetic polymers in which the high molecular weight substance is obtained by a polymerization reaction from one or more simple monomeric substances.

Historically, the derived polymers of cellulose constitute the beginnings of industrial polymer chemistry dating from the birth of celluloid in 1862 and, though this class continues to make an important contribution to industrial polymers as a whole, there are inherent limitations in the potential range of products and properties where one of the principal starting materials is a naturally occurring polymer. Such a limitation is absent in the area of wholly synthetic polymers, the diversity and scale of which have wrought irreversible changes in the chemical industry and, to an increasing degree, is inducing changes in the fabric of western civilization itself.

Synthetic polymers are conveniently divided into two groups:

(i) condensation polymers formed by multiple condensation reactions between chemically reactive functional groups attached to one or more species of reactant. Polyamides of the Nylon type, polyesters

2

such as Terylene and the wide range of urea-formaldehyde and phenol-formaldehyde resins are examples of this type.

(ii) Addition polymers formed by repeated addition reactions between chemically reactive monomers. In this group, the vinyl polymers based on monomers of the type, $CH_2=CHX$, where X may, for example, be a hydrogen atom, a halogen atom, an acetate, phenyl, cyanide or other group capable of conferring the necessary chemical reactivity on the molecule, constitute industrially and commercially the most important class. Of these vinyl polymers, polythene, or polyethylene as it is commonly called in the United States of America and in some other parts of the world, is a limiting type since it is based on the monomer ethylene, $CH_2=CH_2$, in which the substituent X is a hydrogen atom.

In terms of tonnage, world polythene production exceeds that of any other synthetic polymer. In the United States alone, the production figures[1] were 2,700 million pounds in 1964, with projected levels of 3,100 and 4,700 million pounds in 1965 and 1970, respectively. That this scale has been reached in less than thirty years marks the invention and development of polythene as one of the salient achievements in the history of the chemical industry and, for this reason alone, worthy of detailed study in a manner not hitherto attempted. The emphasis is placed on the periods and circumstances that preceded manufacture on an industrial scale and, in particular, the British discovery and development before the Second World War and to a smaller extent the more recent work in Germany and the United States. Many studies in innovation have begun with the implied or stated hypothesis that there exists some common pattern either in the act of invention, or in the process of converting an invention into an innovation. One consequence of this approach has been the tendency to make a rather superficial examination of a large number of cases to try to elicit some kind of pattern or to underline possible common factors. This is not the object of the present work in which the first aim is to trace in considerable detail the scientific background of particular inventions which have led to widely recognized, major innovations.

The industrial prehistory

Polythene was first made as a waxy solid in very small amounts during

---

[1] *Chemical and Engineering News*, *43*, no. 1, p. 12, 4 January 1965.

the period, 24–29 March 1933, in the laboratories of the Alkali Division[1] of Imperial Chemical Industries Limited at Winnington in Cheshire. Its discovery was made within a short distance of the statue of Dr Ludwig Mond who, with his partner, J. T. Brunner, in 1873 began to construct a chemical works on the Winnington site to produce sodium carbonate by the ammonia-soda process. The Brunner, Mond Co. Limited which was formed from this partnership in 1881 was one of the four companies[2] to merge at the end of 1926 to constitute Imperial Chemical Industries Limited. To understand fully the scientific and technological climate that pervaded the Alkali Division of I.C.I. at the time of the discovery of polythene and in the immediately preceding period, it is necessary first to examine the spirit of research and invention which existed in the Brunner, Mond Co.

The contribution of the two partners to the establishment of the works at Winnington were complementary, Brunner on the commercial and financial side and Mond on the scientific and technical. Brunner's contribution, important though it unquestionably was, need not concern us in any detail. On the other hand, Mond's contributions in terms of science and technology were pervading and prodigious. He was of German-Jewish stock, the son of an energetic and strong-willed father and an imaginative mother. Born in 1839, he went in 1855 to study chemistry under Hermann Kolbe at the University of Marburg and proceeded in the following year to the University of Heidelberg where he spent three years under the renowned chemist, Robert Bunsen. Mond left Heidelberg in 1860 without having taken his doctorate and began his industrial career as a chemist in a factory at Mombach near Mainz where acetic acid was made by the dry distillation of wood. From the outset, one of his dominant interests was to recover useful chemical materials from nominally waste products, or to utilize the latter in other chemical manufactures. Two very early examples of this predisposition are afforded by his finding in 1857, while still at Heidelberg, a means of profitably using discarded zinc sulphate and nitric acid in his uncle's electroplating works at Cologne, and while at Mombach of devising a method of making verdigris.

After a period at a Leblanc soda works at Ringenkuhl during which

---

[1] At the time, this was called the Alkali Group. The nomenclature was later changed to Division. The unit is now part of the Mond Division.
[2] The other three companies were British Dyestuffs Corporation Limited, Nobel Industries Limited and United Alkali Co. Limited.

time he began to study the problem of recovering sulphur from the calcium sulphide waste generated by the Leblanc process, Mond moved to Ehrenfeld near Cologne to assist with a process for manufacturing ammonia from waste organic matter. In a room in his uncle's home, where he lived at that time, he continued to work on sulphur recovery from Leblanc waste and in December 1861 patented in France a process for this purpose. Without detailing the chemistry of this process, what was evident and significant at this early stage was Mond's ability to apply scientific principles to the solution of technical problems.

For the next ten years, Mond's activities, centred at the Leblanc works of Hutchinson at Widnes and Smits and de Wolf at Utrecht in Holland, were concerned with the prosecution of his sulphur recovery patent and related improvements which he had developed in these years. This period revealed three aspects of special interest. The first was the extensive experience he gained in devising and supervising the erection and commissioning of plants based on his own patents. The second, and this is highlighted in his negotiations with Hutchinson, was that Mond possessed a well developed sense of monetary rewards appropriate to his inventive and constructive labours. The third feature was his decision taken towards the end of this period, contrary to parental advice, to eschew the safety of a salaried position and, in spite of a lack of capital, to set up in business in England as a chemical manufacturer on his own account.

It is difficult to assess the risks that this decision entailed, but it reveals Mond's confidence in his abilities to succeed both technically and commercially. On his record to that time, such self confidence can hardly be thought of as being misplaced or overestimated. His experience was extensive and varied by the standards of the time; his grasp of chemistry as a science was considerable and his capacity to apply principles to real situations in industrial practice had been successfully tested on more than one problem. His choice of the ammonia-soda process shows a touch of daring, since attempts to work this process in the years between the Dyar-Hemming patent of 1838 and the advent of Solvay et Cie at Couillet near Charleroi in Belgium had been universally unsuccessful. A visit to Solvay in 1872 and a careful study of the plant and of the commercial prospects convinced Mond that a brilliant future awaited the ammonia-soda process. Certainly, the possibility of early profitability is substantiated by a comparison of production costs for Leblanc and Solvay soda made by Mond at that time and fortunately preserved.[1] In spite of this enthusiasm, the agreement he negotiated with Solvay was based not solely on a simple licence arrangement to exploit the Solvay patents in Britain, but also for a complete exchange of the results of future research and development on the process.

Although the plant at Winnington was completed in December 1873, it was not until mid 1874 that it began producing. In the first years of operation there were substantial losses, but by 1876 sufficient profit had been earned to discharge most of the accumulated debts. To a large extent this was achieved by Mond tackling with characteristic energy the manifold initial difficulties often associated with the use of unsuitable equipment and the need to maintain a degree of process control over temperatures, pressures and concentrations then quite new to the chemical industry. By 1881, when the limited company was formed with a capital of £600,000, the programme of expansion and of research had already begun.

As early as 1879, Mond had begun to direct his attention to the salient problems which arose from the operation of the ammonia-soda process. One of these was the need to secure a supply of ammonia for make-up gas and here the approach employed is revealing. All the known processes were examined and as a result a pilot plant based on a proposal of Margueritte and Sourdival was constructed. In this process, nitrogen of the air was fixed as barium cyanide by heating a mixture of barium carbonate and carbon in air, and the resulting product treated with steam to yield ammonia and regenerate the barium carbonate. The project was unsuccessful mainly because of the inability to obtain clay retorts of the required thermal stability in which to carry out the first stage.

A second approach had somewhat unexpected results. In a process patented by Rickmann and Thompson it was claimed that when air and steam were passed through a deep coal fire some of the nitrogen of the air was converted to ammonia. This Mond showed to be untrue, the only nitrogen fixed being derived from the coal. He conceived, however, the possibility of generating by this method a useful power gas for engines and for heating and at the same time recovering about half the nitrogen in the coal as ammonia. Though this process was used at Winnington, its principal development and exploitation was undertaken externally to Brunner, Mond by the Power Gas Corporation and the South Staffordshire Gas Company formed by

[1] J .M. Cohen, *The Life of Ludwig Mond*, Methuen, London, 1956, p. 135.

Mond for this purpose. In the present context, this illustrates an alertness on Mond's part to see and exploit other industrial opportunities.

From the beginning, Mond had introduced a number of direct improvements in the original Solvay process. One of these was the continuous distiller to replace the batch ammonia still of the Solvay scheme; another was in the finishing plant in which the original single stage was changed to a two stage operation. Here again we see convincing evidence of a policy and practice of innovation generated from the early years of the life of the company.

In the beginning, Mond had brought into his employment at Winnington a number of young, German-trained chemists, technicians and other key workers, and while for many years the laboratory and works retained something of a German flavour in chemical science and technology, this was progressively diluted by the recruitment of young men trained in Britain. Initially, there seemed to have been some preference for students of Roscoe at Owens College, Manchester, but subsequently the field was more widely extended. Mond's own training had been a mixture of academic work under Kolbe and Bunsen, both of whom had interests in industrial applications, and a period as an industrial apprentice. He came in later life to emphasize the need for young men to acquire first a high level of training in pure science before turning their attention to practical industrial problems. From these roots, there grew the policy and practice of recruiting to Winnington a steady flow of able, young graduates at a time when this was comparatively rare in Britain.

Although Mond moved his home to London in 1884, he continued until his death in 1909 as a managing director of Brunner, Mond Co. Limited. Until stricken by heart disease in 1902, his contacts with and visits to Winnington were frequent and he continued to influence to a marked degree the activities on that site. In particular, he continued throughout his life to participate in some detail in the alterations and extensions prepared and executed under the immediate supervision of Shellhaas, one of his German speaking assistants. At his home in London, Mond had built a laboratory in which with C. Langer and his elder son, Robert, he continued and expanded his research interests. It was here that the carbonyl process for the recovery of nickel from its ores was discovered and though in itself significant scientifically and industrially, it arose in the first place from attempts to recover chlorine in a usable form from the calcium chloride which was largely a waste product of the ammonia-soda process.

Nearly all of Mond's significant contributions to science and to industry arose in the course of work directed to finding ways of utilizing waste products from one process for other industrial purposes. Something of this approach to industrial research is incorporated in his presidential address to the Society of Chemical Industry in 1889. In his treatment of the theme, 'Necessity is the Mother of Invention' he said:[1]

If this has been the case in the past, I think it is no longer so in our days, since science has made us acquainted with the correlation of forces, teaching us what amount of energy we utilize, and how much we waste in our various methods for attaining certain objects, and indicating to us where, and in what direction, and how far, improvement is possible; and since the increase in our knowledge of the properties of matter enables us to form an opinion before hand as to the substances we have available for obtaining a desired result, we can foresee, in most cases, in what direction progress in technology will move, and in consequence the inventor is now frequently in advance of the wants of his time. He may even create new wants, to my mind a distinct step in the development of human culture. It can then no longer be stated that Necessity is the Mother of Invention; but I think it may truly be said that the steady methodical investigation of natural phenomena is the father of industrial progress.

The material and spiritual bequests of Ludwig Mond to Winnington were manifold. A successful, expanding, technically progressive business provided the secure base of operations; a laboratory established almost at the outset with a potentially fruitful relationship with the industry it served; a policy of recruitment of scientists of high calibre and, above all, a philosophy of research and invention. These were the seeds in the industrial prehistory of polythene, though the period of germination after the turn of the century was yet to come. Before considering this period, it is necessary to trace what had been happening in chemistry, particularly in physical chemistry, at that time.

The scientific prehistory, 1876–1919

The successful operation of the ammonia-soda process depended, among other things, on the careful control of such variables as pressure, temperature and concentration in order that the appropriate solid phases be precipitated and recovered and the continuity of the process maintained. It is not surprising that there had been early

[1]Cohen, op. cit., pp. 181–2.

difficulties in operating the process successfully, since some of the essential physico-chemical principles on which it ultimately rested were largely unknown before 1876. In that year, J. Willard Gibbs published in one of the series of his now famous papers the enunciation of the Phase Rule. This rule prescribes the equilibrium relationships between different phases of the same substance under different conditions of temperature and pressure, and between the phases of a system of several components in terms of such variables as temperature, pressure and concentration.

Its significance was initially ignored or not understood by much of the scientific world, but it was destined to become a central theme in Dutch science for more than a generation. In Amsterdam, Van der Waals from 1877 and Kohnstamm from 1907 and at Leiden, Kamerlingh Onnes from 1882 developed the study of gas-liquid equilibria at high pressures and low temperatures. Van't Hoff, who was Professor of Chemistry at Amsterdam from 1878 to 1895, with Roozeboom and Schreinemakers brought the study of phase equilibria in aqueous-salt systems to a stage which made possible its application to industrial chemical problems.

In Britain, interest in this field was primarily centred at University College, London, under Sir William Ramsay and at Liverpool under F. G. Donnan. Donnan had done postgraduate work at Leipzig under Ostwald who had translated Gibbs' papers into German and had also worked under Van't Hoff, who had moved to Berlin in 1895, before returning to Britain in 1898 as Ramsay's research assistant. His appointment to the Brunner Chair of Physical Chemistry at Liverpool in 1904 had important consequences, for among his early graduates was F. A. Freeth who was destined to play a major role in the early polythene story. When Freeth joined Brunner, Mond in 1907, he brought with him an extensive knowledge and appreciation of the phase rule and its implications and a considerable will to utilize it. At Winnington, Freeth found a fertile field for his researches in phase equilibria, aimed not only at improvements in the ammonia-soda process based on an understanding of the physico-chemical principles involved, but also during the First World War at the development of elegant and effective methods for the manufacture of ammonium nitrate and other explosives. In these activities, successful scientifically, commercially and nationally, Freeth gave the Winnington laboratories a stature in physical chemistry they had not hitherto enjoyed and one which, among other things, facilitated the recruitment of able, young, research-minded graduates. In initiating technical progress based on knowledge gained from research in a field and on systems which bore some demonstrable relationship to the interests of Brunner, Mond, Freeth had in large measure continued the tradition and practised the precept set out by Mond in his presidential address to the Society of Chemical Industry in 1889.

Freeth's role in the background of the polythene story was a significant one. He came to Winnington with a high level of scientific training in an appropriate field which he enthusiastically applied with scientific and technical success. Moreover, he introduced the quantitative approach of the new physical chemistry and recognized the need for highly developed laboratory techniques to obtain the data that this approach demanded. Without Freeth's contribution and stimulus at this time, the laboratories at Winnington may have come to serve no purpose beyond routine process control and empirical trouble-shooting, activities which were scarcely likely to have led to the events after 1919.

The discovery of polythene, 1919–35

In March 1919, Freeth who had been chief chemist at Winnington since 1909 recommended in a report on the future organization and development of the laboratories that there be a small staff employed on detached scientific work covering the more important branches of chemistry. This recommendation contains two somewhat ambiguous phrases, namely, 'detached scientific work' and 'the more important branches of chemistry'. There are many different levels of detachment and even at that time it is doubtful if any two chemists could agree on what were the more important branches of the subject. The particular work which led to the polythene discovery ultimately came under this heading, but it did not begin in that way. Its continuation at critical times and extension to final success were, however, greatly facilitated by the climate of opinion which this report engendered.

Freeth had developed at Winnington considerable expertise in the study of heterogeneous equilibria and, as has been pointed out, held firm views on the need for high quality physico-chemical data and for the techniques used in obtaining them; on the other hand, he had little interest in searching for new products. At that time, the Physics Laboratory at Leiden under Kamerlingh Onnes was outstanding in this field of laboratory techniques. There, it had been recognized that the skilled glassblower and instrument technician had important roles and, as a means of securing this resource, Kamerlingh Onnes had established as part of his laboratory special facilities for the training of such personnel. Moreover, he had for some years under the inspiration of Van der Waals been developing techniques for the measurement of gas isotherms at pressures up to 100 atmospheres and at temperatures as low as that of liquid helium which he had first succeeded in liquefying in 1908. The visit Freeth made to the Leiden laboratory in December 1919 was to have important consequences a year later.

Towards the end of 1920, Freeth was asked to consider the fractionation of coke oven gas as a source of hydrogen for use in the synthesis of ammonia by a high pressure process then being developed at Billingham. The influence of this request on future events was significant, but, before considering the consequences in detail, it is necessary to relate briefly how the Brunner, Mond Co. became involved in the Billingham development. The Haber process for the synthesis of ammonia had been successfully employed in Germany during the First World War, and towards the end of this conflict the British government had decided to build a works for a similar purpose at Billingham. Site preparation had begun when the Armistice was signed and the project was then taken over by the Synthetic Ammonia and Nitrates Co. The shares in this company were owned by Brunner, Mond who had also bought the Billingham site. It was not, therefore, surprising that this particular problem should have been sent to Freeth at Winnington. The synthesis of ammonia had for many years been a central problem in the chemical industry and one to which Mond as early as the 1880's had addressed himself without success. The successful solution of this problem by Haber opened the way to high pressure technology in the chemical industry, though at this stage only to a level of a few hundred atmospheres.

Freeth's reaction to the posing of this problem was to send one of his staff, T. T. H. Verschoyle, to work in Kamerlingh Onnes' laboratory at Leiden in order to acquire experience in the techniques of studying vapour-liquid equilibria which would be relevant to any attempt to separate hydrogen from coke oven gas by fractional distillation. Verschoyle returned in 1922 to begin work on this problem and, following the practice established by Kamerlingh Onnes, Freeth also engaged a glassblower and two instrument makers trained in the Leiden school. When J. C. Swallow joined Brunner, Mond in 1922, he was also sent to Leiden for research experience and upon his return in 1924 was given the task of setting up at Winnington techniques for determining physico-chemical data. This had for a long time been a basic goal in Freeth's plans.

Two new characters now enter upon the scene. R. O. Gibson who had graduated under Donnan in 1924 was recommended by the latter to Kamerlingh Onnes as a promising research student. Gibson arrived in Leiden in 1925 and began work on the measurement of low temperature isotherms of neon. In October of that year, he visited Amsterdam to consult Dr A. Michels on a technical problem of sealing a glass panel into a thermostat tank. Michels who was working under Kohnstamm had in 1919 revived high pressure isotherm work, a field in which Kohnstamm had earlier been engaged as an assistant to Van der Waals. Michels was by nature a perfectionist and being dissatisfied with the accuracy of the measurements made hitherto sought to establish improved methods. From this initial visit of Gibson to Amsterdam, there began an association between these two men on a personal as well as a scientific level which was to have important consequences. This started with weekly visits by Gibson to Amsterdam in preparation for measurements he proposed to make there at the end of the summer of 1926 after completing his work at Leiden. The measurements were duly made, but, because of a calibration error, plans were agreed for Gibson to repeat them in the summer of 1927.

In October 1926 Gibson joined the Winnington laboratory primarily because of the prospect of continuing in the field of work in which his interests had developed. Almost immediately, a new research manager appointed over Freeth proposed to stop the low temperature work which had attracted Gibson to Winnington. By March 1927, however, this decision had been reversed. Freeth had been transferred to London along with W. Rintoul from Nobel Industries Limited as joint research manager for Imperial Chemical Industries Limited which was formed in December 1926. The research manager at Winnington was replaced by Freeth's brother-in-law, H. E. Cocksedge, who had joined Brunner, Mond some eighteen years before. New laboratories at Winnington were also commenced in that year. These moves rescued Gibson's hopes, but, perhaps more importantly, placed Freeth in a central position of some influence within the larger framework of I.C.I.

As arranged, Gibson returned to Amsterdam in his summer holidays of 1927 to repeat with Michels the measurements they had attempted a year earlier. Such was the personal relationship between the two men that Michels journeyed to England to spend his holidays as Gibson's guest. Together, they visited Freeth in London and with his encouragement Billingham as well. The impact of Michels on Freeth and on the senior staff at Billingham was immediate and far reaching. As a result, Freeth and Rintoul visited Michels in Amsterdam, Michels was encouraged to make a further visit to the United Kingdom and by September 1928 Gibson had been seconded to Freeth's staff and sent to work with Michels on the viscosity of compressed gases, a problem which had emanated from Billingham. At the same time, at Freeth's prompting, I.C.I. made a substantial contribution to the Van der Waals' Fund which had been the financial basis of Michels' work. In June 1929 Gibson was joined in Amsterdam by M. W. Perrin who had been appointed by Freeth and attached to his own staff. The connection between Freeth, Winnington and Amsterdam was now firmly established repeating the pattern which had earlier been developed between Winnington and Leiden. In both cases, Freeth had played the role of advocate and manipulator.

Swallow, who in 1924 had set up his section at Winnington for the measurement of physico-chemical data, was not slow to recognize the shift of interest in 1928 towards high pressures and to seek an excuse for incorporating this change of emphasis in his programme. The opportunity came in mid 1929 when the Dyestuffs Group of I.C.I. agreed to finance a programme of work under Swallow on what came to be known as pressure freezing. The object of this work was to see if the effect of pressure on eutectic compositions could be used to effect improved separations in the crystallization of dyestuffs intermediates. By the beginning of 1931, equipment for this project ordered from Michels had been delivered and Gibson was recalled to Winnington to undertake the work, while Perrin remained in Amsterdam, to be joined in October 1930 by E. Hunter and, upon his return a year later, by W. H. Rintoul, the son of W. Rintoul, the joint I.C.I. research manager.

While these arrangements were being made, the whole future of the programme at Winnington was placed in jeopardy when, in July 1930 under the pressure of economic circumstances of that time, a review was made of the research programmes being undertaken. The basic physico-chemical studies including work at elevated pressures survived, though other programmes, notably that known as the coal-oil project aimed at the hydrogenation of coal at 400 atmospheres, was discontinued. Another piece of work which fell by the wayside in the review of 1930 was an investigation by Swallow's group of the separation of hydrogen for re-use in hydrogenation. This was related to the coal-oil project and the decision to discontinue it is not surprising. As a result of the pressure freezing work which yielded no fruitful result and the discontinuance of the coal-oil and associated programmes, there became available by 1931 quite an array of sophisticated apparatus including a vessel for use at 3,000 atmospheres and a mercury-gas compressor. Safety cubicles had also been built in the course of 1929. The investment in experimental facilities and in trained manpower was now considerable. Verschoyle, Swallow, Gibson, Perrin, Hunter and Rintoul (junior) had between them spent more than twelve man-years at Leiden or Amsterdam. Freeth, Rintoul (senior) and Cocksedge were heavily committed and the equity in equipment, facilities and ancillary skills cannot have been other than large by the standards of the time.

Several significant developments emerged in the critical year of 1931 and the early months of 1932. The first was a proposal from the Dyestuffs Group Research Committee, comprising members of the Dyestuffs Group together with Freeth, Rintoul and three consultants, Professors Robinson, Thorpe and Heilbron, for a programme of work on organic chemical reactions at high pressures. The committee had the advantage of consultations with Michels who suggested that pressures of the order of 10,000 atmospheres would be necessary to obtain major effects and of the services of its consultants among whom Robinson, in particular, had a widely recognized flair for predicting the ability of organic chemical reactions to proceed. The second was a report stimulated by Cocksedge and written by Swallow and Perrin (who was still in Amsterdam) in which they recommended that high pressure chemistry in the range 1,000 to 20,000 atmospheres be adopted as a long range project for the Winnington laboratories. Neither of these proposals was in any sense made in isolation. On the one hand, considerable facilities were available for pressures up to 3,000 atmospheres and, on the other, there was growing evidence in the literature notably from the work of Conant and of Bridgman that, in both the organic and inorganic fields, reactions and transformations which did not proceed at low pressures could be induced to do so at very high pressures. Among these was the polymerization of butyraldehyde by Conant.

The reactions proposed by Robinson need not concern us here in any detail, except that they involved binary systems of three types, liquid-liquid, gas-liquid and gas-gas. In a majority of cases in which a gas was involved, the preferred component was carbon monoxide or ethylene, while Robinson's central suggestion led in the limit to the reaction between these two gases themselves.

Among the recommendations of the Swallow-Perrin report was the reasonable proposal that if work on organic reactions at high pressures were to be undertaken, an organic chemist from Dyestuffs Group should be attached to the Winnington team. The choice fell on E. W. Fawcett who had been a member of the coal-oil group which was disbanded during the cuts of 1930. Fawcett had visited North America and, significantly, had spent some time with Carothers' team at du Pont.[1] Polymeric substances were, therefore, no new thing for him, though Carothers' work had not at this time reached full flower.

The programme on chemical reactions at high pressures began under Gibson and Fawcett in February 1932 using pressures up to 3,000 atmospheres and temperatures up to about 200°C. During 1932, some *ad hoc* experiments were made on the reaction between carbon monoxide and ethylene which produced not acrolein, as had been suggested by Robinson, but a polymer of acrolein. In December of that year, progress was reviewed and the Dyestuffs Committee recommended that attention be directed to five gas-liquid systems involving carbon monoxide and to the ethylene-benzaldehyde reaction. During this period, the laboratory originally provided for the coal-oil project was being refitted by the research engineer, B.E.A. Vigers, for the work on these reactions and by the beginning of 1933 a temporary installation was ready. The experimental rig now included an hydraulic test pump which, while affording a more convenient method of raising the pressure than that employed hitherto, was limited to 2,000 atmospheres. It was on this rig that the experiments of March 1933 were carried out.

The critical experiment began on Friday 24 March 1933; in this the reaction between ethylene and benzaldehyde was attempted at 2,000 atmospheres at 170°C. The pressure held well over the night of 24–25 March, but a leak developed during the week-end. On Monday

[1] M. Kaufman in *The First Century of Plastics*, The Plastics Institute, 1963, p. 82, states that Fawcett had spent a couple of years in the United States and had worked for some months with Carothers' team. This appears to be an overstatement.

27 March there was virtually no pressure in the apparatus and, on dismantling, it was found that the benzaldehyde had been blown out to be mixed with the oil, but the steel gas inlet tube in the apparatus was thinly coated with a waxy solid as though it had been dipped in paraffin wax. This was the first recorded observation of the formation of polythene. On 29 March the experiment was repeated and this time, with all the fallibility that experimentalists are heir to, they had forgotten to close the gas stop valve when initially attempting to raise the pressure, thereby pushing the benzaldehyde out of the reaction vessel. The experiment was continued and again a small amount of waxy solid was produced. By July 1933, half a gram of material had been obtained and this Fawcett was able to identify positively as a high molecular weight polymer of ethylene. In the course of these early experiments, unexplained decompositions to yield carbon and hydrogen occurred. While these decompositions were the source of much difficulty and uncertainty, the eventual elucidation of their cause yielded some useful information on the nature of the polymerization process.

This discovery does not appear to have created much interest at that time. The report made by Gibson and Fawcett for March 1933 is, to say the least, prosaic. It reads:[1]

The reaction between ethylene and benzaldehyde has been studied at 2,000 atmospheres and 170°C. A waxy solid which appears to be polymer of ethylene was formed. In one experiment the ethylene decomposed to carbon and hydrogen.

A few weeks later, at the end of April, an interim report was issued in which the melting point of the polymer is recorded as 113° and the incidence of the decomposition to carbon and hydrogen highlighted to the extent that the authors reported:[2]

This decomposition is being further investigated in order to ascertain whether the study of reactions involving ethylene is feasible under these pressures.

Apart from the problem of unexplained decompositions, there were other factors which decreased the impact of this discovery. The waxy solid produced was quite unlike other synthetic polymers such

[1]R.O. Gibson, *The Discovery of Polythene*, Royal Institute of Chemistry Lecture Series, 1964, no. 1, p. 17.
[2]Gibson, loc. cit., p. 20.

as polystyrene and polymethylmethacrylate familiar at that time; only very small amounts were made; the material was very different from the low molecular weight polymers of ethylene in the form of liquids then known; and, perhaps of special relevance, the climate of chemical thought at that time, particularly among organic chemists, was that the only solid materials, the purity and identity of which could be guaranteed, were crystalline and capable of being re-crystallized without change in physical or chemical properties.

In Gibson's account of the period that followed, he suggests that Fawcett had a hunch that the novel type of polymer they had made might turn out to be important. Here we may recall that Fawcett had visited Carothers' laboratory at du Pont, a circumstance which may have sharpened his hunches. In the light of subsequent happenings, however, it is difficult to substantiate Gibson's suggestion. For one thing, Gibson and Fawcett published[1] their work on liquid phase reactions in 1934 and Fawcett who was planning to attend the Faraday Society Discussion on 'The Phenomena of Polymerization and Condensation' held at Cambridge in late September 1935 obtained permission to make a suitable contribution on the ethylene work at an appropriate opportunity. This contribution which summarizes the state of knowledge at that time—some eighteen months after the initial discovery—strongly suggests that no commercial possibility had been recognized. Moreover, this contribution goes close to making public disclosure as the following quotation[2] shows:

In connexion with Professor Staudinger's remarks on the relative ease of polymerization of various unsaturated compounds, I should like to report some preliminary results on the polymerization of ethylene under high pressures. The polymerization of ethylene when carried out at ordinary or moderately high pressures with or without the addition of catalysts, does not usually lead to products of very high molecular weights—the products generally being liquids of molecular weight of the order 100–500. When, however, ethylene is heated to 170°C under a pressure of about 1,000 atm. a slow polymerization occurs leading to the production of a white solid polymer. This is insoluble in acetone, and moderately soluble in benzene, has a carbon and hydrogen content corresponding to the formula $(CH_2)_n$, and a molecular weight by boiling point elevation in benzene of about 4,000. In spite of the fact that the reaction is by no means complete—only about 10 per cent of the ethylene had reacted—no liquid products of low molecular weight were produced, and the reaction appears to be quite analagous to the well known chain polymerization of styrene and similar

[1] E.W. Fawcett and R.O. Gibson, *J. Chemical Soc.*, 384, 1934; 396, 1934.
[2] E.W. Fawcett, *Trans. Faraday Soc.*, 32, 119, 1936.

substances. When an attempt was made to carry out the reaction at a pressure greater than 1,400 atm., a very violent exothermic reaction occurred with the production of carbon, hydrogen and simple hydrocarbons.

Because this contribution goes so close to disclosing the essential features, bar one, of the principal patent,[1] and because permission to disclose the results of the ethylene work had been given—in what detailed terms we do not now know—it is difficult to escape the conclusion that no potential commercial possibilities were seriously envisaged up to late September 1935. It is, however, possible that Fawcett's statement was more extensive than had been intended since it followed immediately after a contribution by Professor Mark in which he claimed categorically that ethylene did not polymerize. That Fawcett clearly knew at that time the importance of peroxide catalysts in vinyl polymerizations is without doubt, since his earliest approach to unravelling the problem of decompositions was to seek to eliminate peroxy components from the benzaldehyde which had initially been present in the reaction vessel.

The effect within I.C.I. also reveals little recognition of the initial discovery, for at the end of 1933 the Dyestuffs Committee which had sponsored the work on chemical reactions at high pressures discontinued its support. After that point, the direction of the work changed from *ad hoc* studies to more systematic physico-chemical work. Fawcett left the high pressure team for other duties in 1934 and in the middle of that year Perrin was recalled from Amsterdam to join Gibson. In October 1935 Perrin was put in charge of the team and Gibson transferred at his own request to another section at Winnington. These changes taken together do not in themselves suggest any immediate or concerted effort to overcome the two outstanding problems—the quantity of product and decompositions—nor do they show evidence of any appreciation of the potential of the new high polymer.

In the years 1933 to 1935, there had been considerable technical improvements in the experimental vessels and ancillary equipment largely resulting from designs due to W. R. D. Manning, the research engineer, and aimed, among other things, at improving the safety of work at these elevated pressures on which the Winnington management had insisted. Michels also contributed some of the detailed design ideas, but it is significant that in this period, in particular, the

[1]B.P. 471,490, Improvements in or relating to the polymerization of ethylene.

3

technical improvements were being generated predominantly at Winnington itself. These improvements which were completed by December 1935 gave to the Winnington laboratories an experimental facility that was unique.

It was natural that Perrin who, on taking over the group, had decided to begin by again examining the effects of high pressure on ethylene, should discuss with Manning the vital problem of decompositions. The team by this time had grown and now comprised Perrin, Manning, E. G. Williams who had joined the group in September 1934 and J. G. Paton who had been added in September 1935. The first experiment was done by Perrin and Manning in the evening of 19 December 1935, but no written record of it exists. It appears that they intended to compress the ethylene to 2,000 atmospheres at 170°C, but owing to a leak they had difficulty in attaining this pressure and as a result continued to add ethylene. On cooling the autoclave and lowering the pressure in that order, it was found to be full of a white powdery polymer. Later, it was established that if the pressure was released first and the autoclave then cooled, the waxy polymer of Gibson and Fawcett was produced. On 20 December 1935, the first 'official' experiment was carried out by Williams and Paton. Again, a good yield of polymer was obtained, but attempts to repeat the experiment evidently led to decompositions.

By January 1936, sufficient polymer had been made to enable some of its properties to be evaluated. Among these were its moulding ability, a volume resistivity of the moulded polymer in the range of good dielectrics, and a film forming capacity to yield a tough, strong, thin, transparent product. Some other important facts had also been established, notably that the decompositions resulted from exothermic polymerization taking place so fast that the temperature became uncontrolled, and that the molecular weight of the polymer increased with increasing pressures. Once sufficient polymer had been made for the properties such as those noted above to be evaluated, it became evident that the material was of potential technical interest.

The original patent, British Patent 471,590 was filed on 21 August 1936 and granted on 6 September 1937. The document bears the names of Fawcett, Gibson, Paton, Perrin and Williams and in accordance with normal practice was assigned to I.C.I. This patent is interesting from a number of points of view. The first is that it does not include the name of Manning who had contributed much to the apparatus design and construction over a considerable period and was personally involved in the experiment with Perrin on 19 December 1935. The second point is that it sets out certain ranges of temperature, pressure and oxygen concentration to be used in order to obtain acceptable polymers and to avoid decompositions. Thirdly, it lists a range of applications. Other patents[1] followed in which the polymerization conditions were progressively refined. In these the names of Paton, Perrin and Williams alone occur. By this time the important role of oxygen as a radical initiator had been more fully appreciated and, though the comparative uncertainty of initiation in this way has long since been replaced for most grades by a specifically added, deliberately designed radical generation system, decompositions still occasionally occur. No longer were the few scraps of polymer made by Fawcett and Gibson of academic interest only; the period of discovery and rediscovery was over, the uncertainties of development were about to begin.

The development of polythene manufacture, 1936–9

It is instructive to begin this section by summarizing the state of the art at the time the first decisions had to be made. A new polymeric material had been discovered and made on a 10 gram scale, a simple evaluation of its more important properties carried out, and some potential general uses had been noted. It differed from most other polymers known at that time and had been made by techniques and under conditions which, even in the laboratory, would be considered unusual and with which it was not always possible to achieve consistent results. The molecular weights of the polymers made at this time were still quite low on the scale of commercial polythenes today and there was a limited amount of knowledge on how the conditions of pressure, temperature and oxygen concentration could be manipulated to obtain products to meet a specification if one could have been prescribed. There were no particular uses in the sense that this material had not been invented to meet a prerecognized need and the task of making a market survey, much beloved by present day product developers, presented difficulties of formidable magnitude. Moreover, this discovery had been brought forth in an area of the chemical industry which for generations had been concerned with the manufacture and supply, often under long-term bulk contracts, of basic inorganic chemicals, a field and a tradition far removed from

[1] U.S. Patents 2,153,553 and 2,188,465.

the ultimate consumers of plastics as they then existed. At the time, I.C.I. had no special interest in ethylene as a raw material. In brief, the outlook for a child so born must have seemed bleak indeed.

The first requirement in development was to make sufficient material for a more widely based evaluation of its properties. Initially, this was done by increasing the reactor size from about 80 ml to 750 ml to produce material on a pound scale and then, early in 1937, from 750 ml to 9 litres. The second requirement was to procure a means of mechanically compressing ethylene to the required pressure. This unit, designed by Michels, also became available early in 1937 along with the 9 litre autoclave. The third need which became imperative as the scale increased was to learn how to control the polymerization by adequate means of controlling the pressure, temperature and oxygen concentration and to devise a means of removing the high heat of polymerization. The solution of these problems and the need to operate continuously led to the erection on a works' site of a small plant to produce ethylene from ethyl alcohol and the transfer of the high pressure equipment from the confines of the laboratory. This experimental plant came into existence in March 1938, and by the end of that year a ton of polythene had been produced and some useful experience obtained on operating a small production unit.

Simultaneous with these developments and as increasing quantities of polythene became available in the period 1936–8, attempts were made to find specific possible uses. Metropolitan-Vickers Limited carried out a detailed examination of its electrical properties and other groups in I.C.I. were acquainted with the material. Plastics Group was given the responsibility of examining possible moulding applications, a field in which they had expertise, but other possibilities were the responsibility of Alkali Group at Winnington. The most promising leads came from the Telegraph Construction and Maintenance Company in the person of J. N. Dean and from B. J. Habgood who, before joining Dyestuffs Group of I.C.I., had had first hand experience of the cable industry. Gutta percha and modified gutta percha had been used since 1856 as a cable insulant, but because of the introduction of higher transmission frequencies there had arisen a need for an insulating material with a lower electrical power loss. Moreover, gutta percha was a jungle and not a plantation product and, therefore, erratic in supply. By September 1938, Dean's company had tested a mile length of polythene insulated cable laid between the Isle of Wight and England. The polythene used in this test had been produced on the laboratory plant and, though extrusion presented problems arising primarily from the fact that the melting point of polythene was some 60°C higher than gutta percha, the test was successful. This led to the placing in September 1938 of an order for 100 tons of polythene for delivery by the middle of 1939.

The first commercial plant built at Wallerscote initially to meet this requirement for the submarine cable industry came into operation in September 1939, prophetically on the day on which Germany invaded Poland. The plant was equipped with fifty litre vessels and, apart from the chemical problems of supplying ethylene of the required purity, much depended on solving the engineering problem of compressing the gas to the required pressure. The I.C.I. group of companies had by this time considerable experience in ammonia synthesis and other applications of high pressure technology, but only up to a few hundred atmospheres. To achieve the higher pressures required was a task beyond the engineering industry in Britain at that time and in consequence a machine of Michels' design, originally intended for laboratory service only, had to be scaled up and converted to continuous operation. In this machine, gas on one side of U-tubes fitted with mercury seals was compressed and displaced over the mercury by forcing in oil from hydraulic intensifiers on the other side. Hunter's[1] comment on this machine, recorded in 1961, reveals the intensity of the problem in the following words:

These machines were condemned in principle by all engineers who saw them. Apart from difficulties in manufacture, they required standards of cleanliness and constant attention beyond anything previously experienced in chemical plant machinery, but they had to be persisted with in the absence of anything better. The engineers eventually overcame the problems of designing more conventional compressors for the high pressures, and the last of the original designs ceased work in 1950. But it is worth stressing that the development of polythene . . . depended on a highly unconventional piece of laboratory equipment designed by a physicist for use by physical chemists.

One aspect of this phase is the financial basis on which the decision to build the first commercial plant at Wallerscote was taken. While polythene in other physical forms, for example film and solvent spun fibre, had been made in experimental amounts in 1938, the only firm market appears to have been the submarine cable industry which

[1] E. Hunter, *Advancement of Science*, *18*, 171, 1961.

could scarcely be described as one with a large growth potential. Hunter[1] recalls that at the time the estimated demand was 2,000 tons per year. The technology was new in an extreme form as evidenced by the compressor story noted above; the research and development charges extending over a number of years cannot have been inconsiderable; the costing, at best, was uncertain and the proposed price of five shillings a pound probably as experimental as the plant and product itself. There have been several points in this narrative at which the work in research and development might have been justifiably terminated, and yet this was not done. Seldom can there have been so little commercial encouragement arising from an estimated return on the investment or from any of the more or less sophisticated indices useful in arriving at an investment decision, but these considerations, if indeed they were seriously contemplated, were to be summarily removed by the demands of war.

Polythene in the Second World War

The polythene made in the Wallerscote plant was destined to be used not for submarine cables, but to meet a need for a flexible, high frequency cable for airborne and ground radar. The significance of polythene in this field is perhaps best told in the words[2] of Sir Robert Watson Watt, the inventor of radar:

The availability of polythene transformed the design, production, installation and maintenance of airborne radar from the almost insoluble to the comfortably manageable. Polythene combined four most valuable properties in a manner then unique. It had a high dielectric strength, it had a very low loss factor even at centimetric wavelengths, it could fairly be described as moisture repellant, and it could be moulded in such a way that it supported aerial rods directly on watertight, vibration proof joints backed up by a surface on which moisture films did not remain conductive. And it permitted the construction of flexible very high frequency cables very convenient in use. A whole range of aerial and feeder designs otherwise unattainable was made possible, a whole crop of intolerable air maintenance problems was removed. And so polythene played an indispensable part in the long series of victories in the air, on the sea and on land, which were made possible by radar.

The 1939 plant soon proved inadequate for the demands and a

[1] E. Hunter, personal communication, 1965.
[2] Quoted by J.C. Swallow in *Polythene*, ed. A. Renfrew and P. Morgan, Iliffe & Sons, London, 1957, p. 7.

larger plant was erected. Although control of the process improved as experience grew and methods of assessment, especially that of product grading in terms of the melt viscosity at constant rate of shear (melt flow index), the plant extensions in those years were directed almost entirely to increasing the scale without otherwise altering the process. Cost was not, in itself, a major consideration and for a long time exceeded the price that had been set.

In 1941, information on polythene was communicated to the United States apparently through two channels. Some general information about its use in radar cables was made available as part of the deal by means of which certain ships and war materials were supplied to Britain, but a more important channel was made possible by the I.C.I.-du Pont Technical Agreements. Sir Harry McGowan of I.C.I., recognizing the signal importance of polythene to the Allied war effort, offered through Mr Walter Carpenter of du Pont to receive an investigating team at Winnington. This offer was accepted and the du Pont group visited the United Kingdom in October and November 1941. They took back to the United States complete information including all the engineering drawings, specifications and operating instructions for the plants at Wallerscote and Winnington.

In the event, du Pont preferred to develop a different kind of plant involving a wet tube in place of the I.C.I. autoclave, largely because of doubts about dissipating the high heat of polymerization.[1] Independently, and without specific information from I.C.I. or du Pont, the Union Carbide and Carbon Corporation developed a dry tube process. These two plants, built in West Virginia with the assistance of the U.S. government, were in production by 1943. By the end of the war, output in the United States exceeded that of about 1,500 tons per annum in Britain.

One of the problems that beset the British manufacturers was that extrusion equipment available for cable manufacture was unable to handle satisfactorily the grade of polythene being made for this purpose. This difficulty was alleviated by plasticizing the polythene with $12\frac{1}{2}$–15 per cent of polyisobutylene. In the United States, where extrusion technology was more advanced, this proved unnecessary and the higher molecular weight polythenes could be successfully processed.

[1] J.C. Swallow, *Council News*, Plastics Division Council, I.C.I. Plastics Division, April 1958.

The removal of wartime restrictions led to an immediate expansion in production capacity, especially by Union Carbide, and material now became available for development of other uses. The vigorous consumer-oriented approach in North America coupled with the size of the market, the more favourable, competitive economic climate than in Europe at that time, and the established expertise in plastics manufacture and utilization led to rapid expansion, especially into the film and injection moulding fields. The special suitability of polythene for these purposes soon became apparent and since 1954, when the earlier shortage of material began to be overtaken by production capacity, the expansion has been astonishingly broad and rapid. Two factors are of special note in this period. The first, in 1952, was a court decision in a case brought by the U.S. Government against I.C.I.-du Pont under the Sherman Anti-Trust Act, the result of which was that I.C.I. was required to offer licences for polythene manufacture to other companies in the United States. Among those who took up licences were Texas Eastman Company, Monsanto, Spencer, National Petrochemicals, Dow, and Koppers. A number of these licencees also purchased the I.C.I. know-how on the use of autoclaves. The second feature of the post-war period was the decline in ethylene costs, particularly in North America. Expanding, competitive production facilities and decreasing raw material costs launched polythene into many applications and on a scale beyond the wildest dreams of those by whose faith the first commercial plant of a few hundred tons per year capacity was established at Wallerscote in 1939. In all this, none had played a more significant role than Dr E. Hunter who was directly concerned with the I.C.I. programme from 1936 to 1958 and who, in J. C. Swallow's words,[1] 'probably knows more about the fundamentals of making polythene than anyone else in the world.'

High density polythene

Polythene made by the I.C.I. high pressure process in the fifteen years which followed the first commercial production in 1939 had, in round figures, a density of $0 \cdot 92$ g per cm^3 and a number average molecular weight in the range, 20,000–40,000. As early as 1939, Bunn[2] had shown that the polymer was not entirely crystalline as might have been supposed from the symmetrical molecular character

[1] Swallow, op. cit.
[2] C.W. Bunn, *Trans. Faraday Soc.* 35, 482, 1939.

of the monomer. The crystalline regions, or crystallites, were shown to be less than 100 Å in diameter separated by amorphous regions, it being supposed that the whole mass was tied together by the long linear chains of more than 1,000 carbon atoms passing through a succession of crystalline and amorphous regions. This proved to be an over-simplified picture and the polythene made by the I.C.I. process did not in fact comprise wholly linear molecules, but ones in which there were some short and long branches. An early result of these and related findings was the recognition that if the extent of branching could be varied, materials with different degrees of crystallinity and hence different densities and other physical properties, notably flexibility and extensibility, should result. Until the middle 1950's, however, the task of supplying the rapidly growing market with polymer of the conventional type fully occupied the producers.

In the early 1950's a new wave of invention appeared, when three new processes for making polythene were disclosed; all of these were operable at low pressures and yielded polymer of higher density mainly in the range, 0·94 to 0·96 g per cm³. Of them, that of Karl Ziegler of the Max Planck Institute for Coal Research at Mülheim, West Germany, proved to be of the greatest significance. The other two were by the Phillips Petroleum Co. and Standard Oil of Indiana, both United States companies long established in the petroleum products field. All three processes employed distinctive catalysts quite different from the free radical initiators used in the I.C.I. high pressure process.

The Ziegler process employs, typically, a catalyst formed by reacting aluminium triethyl with titanium tetrachloride in a hydrocarbon solvent in the absence of oxygen and water. The coloured product so formed will polymerize ethylene spontaneously at atmospheric pressure and room temperature, though somewhat elevated conditions are often preferred. The resulting products first made by this method were of such high molecular weight that they were difficult to process on conventional machinery, but by varying the proportions of the metal alkyl and transition metal compound, polymers of molecular weight suitable for processing could be obtained. The question which poses itself is how Ziegler came upon this particular group of catalysts which, by most standards, are rather esoteric. To attempt to answer this question, it is necessary to examine Ziegler's research activities in the preceding years.

After taking his doctorate at Marburg, Ziegler taught and did research in several German universities for about twenty years before 1943 when he was invited to become Director of the Max Planck Institute for Coal Research at Mülheim. Such an invitation was a considerable honour to any scientist in Germany even in those unhappy years. Ziegler has later recorded[1] his doubts at that time whether he could successfully adapt himself from his long period in academic work to the environment of an applied research institute concerned with coal. He accepted the invitation on condition that he be permitted the 'freedom to work on any problem of organic chemistry without the obligation that the problem should have a close relation to coal, or any relation at all.'

On its face value, this comment suggests that Ziegler looked upon himself as a detached academic researcher eschewing any element of practical use that might flow from his work. An examination of his published work during the twenty years preceding his appointment at Mülheim suggests that this may be a distorted picture. In this period, about a hundred publications bearing his name are listed in Chemical Abstracts and among them are to be found nearly twenty patents which are brought together in Appendix I. It is evident that Ziegler was, from the beginning of his career, conscious of the possible commercial implications of his scientific research.

Ziegler's published research work is extensive and varied, but the central theme over a long period was his interest in the chemistry of metal-alkyls, initially alkyls of the alkali metals, especially lithium. As early as 1939, he had become involved in work on synthetic rubbers based on butadiene in which polymerization was effected by a lithium alkyl. It is almost certain that his interest in the aluminium alkyls developed from the work on the lithium compounds and, in retrospect, a fairly obvious extension of it. We have seen this pattern before in the evolution of Mond's interests and in the broadening of the phase equilibria work of Freeth at Winnington. Nor is there anything strange about this, for it is the way in which most scientific research proceeds and one which is almost second nature to research workers in science generally and in chemistry in particular.

In his 1952 paper,[2] Ziegler reported the effectiveness of aluminium trialkyls as catalysts for the polymerization of ethylene and not only offered an explanation of the reaction, but also commented on the

[1] K. Ziegler, *Research in the Max Planck Society*, Shell Chemical Co., 1962.
[2] K. Ziegler, *Angewandte Chemie*, 64, 323, 1952.

possible commercial significance. Among the products described were those containing 16 to 18 carbon atoms in the group of important fatty acids. At this point in time, the possibility of synthesizing a conventional polythene by extending this process to a thousand or so carbon atoms seems to have arisen. It was well known at that time that growth reactions of this kind could be severely limited by traces of impurities, and a systematic attempt was therefore made to track down possible inhibitors by adding potential offending substances one at a time to the reaction system. In the course of this study, it was found that some substances, far from being inhibitors, greatly accelerated the reaction and were capable of yielding a polymer of almost any size. This work on which the master patents[1] rest is described in some detail in papers[2] published in 1955.

The development of a process based on this discovery posed a number of problems. Apart from the question of maintaining the reaction system free of water and oxygen, there remained that of freeing the ultimate product from its metallic component. Unless the latter could be successfully removed or rendered inert, the electrical properties of the product would be severely impaired. To some extent this has remained a problem, but since many of the uses of high density polythene made by the Ziegler process have been in areas where this requirement is not of special importance, and because in some applications the product is superior to low density polythene, its commercial exploitation has followed rapidly.

Ziegler's discovery had, however, implications more far reaching than the extension of the range of polythenes. It was quite natural that chemists working with ethylene polymers should consider an extension to polymers of propylene, the next higher member in the olefine series. Attempts to produce polypropylene by a simple adaptation of the high pressure process had yielded only syrupy liquids or rubbery solids of minor commercial interest. In 1954, G. Natta of the Milan Polytechnic succeeded in polymerizing propylene to a solid high molecular weight polymer using Ziegler-type catalysts. This discovery was more than a simple extension of Ziegler's work, since Natta had recognized that the spacial arrangement of the component groups of atoms in the polymer of propylene would have a profound effect on the physical properties. The propylene molecule

$$CH_3-CH=CH_2$$

[1] Belgian Patents, 504, 160 (21.6.50) and 533, 362 (23.12.53).
[2] *Angewandte Chemie*, 67, 541, 1955; *Kunststoffe*, 11, 506, 1955.

differs from ethylene, $CH_2=CH_2$, in that it is no longer symmetrical, one hydrogen atom of the latter having been replaced by a methyl, CH_3, group. When the molecules join together to form a polymer, three spacial configurations, two ordered and the other disordered with respect to these methyl groups, are possible. The ordered configurations are illustrated schematically opposite.

The disordered atactic material is a soft, rubbery product of unsatisfactory mechanical strength. On the other hand, Natta predicted that isotactic polypropylene, in which the packing density of the chains should be greater, would be a much more useful material. This turned out to be the case; the isotactic polymer was a solid thermoplastic with properties which were in some important respects an improvement on those of polythene. What was of critical importance was the ability of the Ziegler-type catalysts to produce these desirable stereospecific polymers. A further result was that this discovery made possible the copolymerization of ethylene and propylene from which a range of materials with desired physical properties could be made. Syndiotactic polymers have been recognized in nature in the case of polyisoprene and are known to have distinctive properties, but this stereospecific structure has not yet been realized industrially.

Natta, to some extent in contrast with Ziegler, had specific industrial interests. He had, for example, before 1950 produced new synthetic methods for manufacturing methanol and formaldehyde and, though holding an institutional appointment, had for thirty years been a consultant to Montecatini, a large Italian chemical company. The separate but related achievements of Ziegler and Natta were recognized internationally in 1963 by the joint award of the Nobel prize in chemistry.

Polythenes of higher density and crystallinity also arose in the 1950's from discoveries of a quite different kind made independently by Phillips Petroleum and Standard Oil of Indiana. Both were low pressure processes operating at about 1,000 lb per in^2 and both employed solid catalysts. Those listed in the Standard Oil patent[1] were of two types: (i) the metals nickel or cobalt supported on charcoal; and (ii) the oxides of the transition metals of Groups VA and VIA of the periodic system supported on alumina, titania or zirconia. The Phillips patent[2] lists as the preferred catalyst chromium oxide in the sub-hexavalent state supported on steam activated silica-alumina. In

[1] Belgian Patent, 525, 025 (28.4.51).
[2] Australian Patent Application, 864/54 (8.6.54).

CH_3 CH_3 CH_3 CH_3 CH_3 CH_3
 | | | | | |
$-CH-CH_2 -CH-CH_2 -CH-CH_2 -CH-CH_2 -CH-CH_2 -CH-$

(a) ordered, with all CH_3 groups on same side of the plane of the carbon-carbon chain (isotactic)

CH_3 CH_3 CH_3
 | | |
$-CH-CH_2 -CH-CH_2 -CH-CH_2 -CH-CH_2 -CH-CH_2 -CH-$
          | | |
        CH_3 CH_3 CH_3

(b) ordered, with the CH_3 groups alternately on opposite sides of the plane of the carbon-carbon chain (syndiotactic).

CH_3 CH_3
 | |
$-CH-CH_2 -CH-CH_2 -CH-CH_2 -CH-CH_2 -CH-CH_2 -CH-CH_2-$
          | | |
        CH_3 CH_3 CH_3

(c) disordered, with the CH_3 groups randomly distributed with respect to the plane of the carbon-carbon chain (atactic).

both cases, the discoveries on which these patents are based appear to have resulted not from work directed to the polymerization of ethylene as such, but from a long standing aim of the petroleum industry to upgrade or modify less valuable streams to products suitable for use in gasoline or other quality fuels. By the time this discovery was made, the petroleum companies especially in the United States were becoming heavily involved in the chemical industry as suppliers of gases such as ethylene for use in chemical synthesis and polymerization. There can be little doubt that there was full awareness of the possible fruits that might grow from new methods of producing polythenes.

These developments taken together made it imperative that I.C.I. should seek to protect its special position as the inventor of the high pressure process. By 1954 this company, too, had discovered how to make polythenes of higher density (up to 0·94) by modifying their original process. While the high pressure process in autoclaves or tubular reactors continues to be the method of most extensive use, low pressure processes, particularly that based on Ziegler catalysts, are now responsible especially in Germany and the United States for a substantial tonnage of polymer. In retrospect, one can scarcely fail to admire a certain elegance of the Ziegler process with which both ethylene and propylene may be polymerized to yield homopolymers and copolymers; with propylene, stereospecific polymerization is possible; and with either highly crystalline, ordered polymers are the normal result. In practice, however, the choice of process for making polythene is primarily determined by the properties required in the product.

Commentary and discussion

In the preceding pages, the aim has been to set down in some detail the factual situation concerning the discovery and early development of polythene. We now turn to an examination of this record from the point of view of a study in innovation. Some incidental comments have been included in the narrative where these appeared necessary or desirable in order to underline or amplify facts or statements and to stimulate some of the more general questions which are taken up in this section.

One of the first of these questions is whether the industrial prehistory surrounding Mond is significant in the discovery of polythene.

It may be argued that Mond ceased to live at Winnington in 1884, and was not closely associated with the site after his health began to fail in 1902. His material contribution to the Winnington works and its laboratory from 1873 to 1884 is beyond doubt. After that time, his principal detailed interests at Winnington lay in the area of plant renewal, development and extension. If Mond's contribution to polythene was at most tenuous in a direct concrete sense, it is difficult to deny that the discovery was nurtured in embryo in a climate which owed much to his initial creation. His influence at Winnington was deeply rooted and widespread from management to factory floor; he was a dominant figure technically and personally with sufficient idiosyncracies in outlook, method, manner and dress to have cast a lasting impression; there was a notable consistency in his industrial outlook, particularly as regards research and invention in industry. But above all, he had an appreciation unusual for his time of the gains that can accrue from the application of scientific principles to the solution of practical problems. In this regard, the course of his several enterprises was set not by empiricism alone, but by the consistent striving to apply the principles of chemical theory.

A feature of Mond's life was not so much his separate achievements as a scientist, inventor or industrialist, but in the way in which these three roles were integrated and continued in this form throughout his whole life, certainly to a much later age than would appear to be common today. He undoubtedly enjoyed his wealth, but this did not diminish in him the independent spirit of enquiry as the history of the discovery, development and industrialization of the carbonyl process amply demonstrates. Possibly because he spanned within himself such a range of achievements, there was in many new developments with which he was associated a tendency to proceed rather precipitately from a laboratory preparation to an industrial scale. This necessarily brought with it expensive problems of doing the chemical engineering along the way, a feature which, for rather different reasons, was also part of the fabric of the first commercial production of polythene.

Whatever may be the detailed assessment of Mond's contribution to the climate that existed at Winnington at the turn of the century, it undoubtedly created one which Freeth found favourable. It would be unrealistic to suggest that Freeth met no difficulties or suffered no hindrance, but he was not one to be diverted easily from a path once chosen. His part in the discovery of polythene was peculiarly vital.

In his early years at Winnington, he clearly practised the Mond philosophy and by 1919 had established himself in a strong, though by no means impregnable, position. His standing as a scientist in his own right, confirmed by his work during the First World War, made it possible for him to promote vigorously a point of view which, while not inconsistent with the industrial framework of Brunner, Mond, was at the same time in tune with the trend of chemistry at large in that period.

It has often been suggested that many of the advances in science and technology have been initiated in the areas between the circumscribed disciplines, or have been made by people who have embraced to a greater or less degree more than one of the conventionally defined fields of science. Physical chemistry is perhaps the oldest of these interdisciplinary fields and one of the first to be blessed with a separate name. Definitions of its scope vary from 'anything interesting' to 'the application of physical and mathematical methods to problems of chemistry', while physical chemists are or were wont to describe themselves as 'chemists who blew their own glass and solved their own equations'. The salient point here is that Freeth, from the beginning of his time at Winnington, clearly recognized the relevance and importance of the quantitative approach of physical chemistry to research and development in the chemical industry. This realization was a theme of his own activities and of the policies he was later to pursue in I.C.I. In our present context, it was the link with which the relationships between Winnington and Leiden and Amsterdam were established.

Much has been made of the recommendation of the Freeth report of 1919 concerning the desirability of a portion of the research effort at Winnington being devoted to detached work. Before considering this aspect in detail, it is necessary first to try to decide more exactly what was intended. The section established by Swallow in 1924 for the purpose of generating high quality physico-chemical data was at the time an essential requirement if the quantitative approach of physical chemistry was to be successfully employed. Such work, in so far as it did not stem directly from individual, specific problems arising from an existing plant, could quite properly be described as detached. At the other end of the scale, research work at Winnington on, say, the influence of sea temperature on the vitamin C content of whales' milk would unquestionably be described as detached. What was most probably in Freeth's mind was the concept of broadly based studies which did not have narrowly defined specific goals and did not have to be justified in detail against existing or known future process commitments of the company.

This was not by any means a new idea, though it may well have been unusual in the United Kingdom at that time. The practice of intensively researching a broad field in which the operating company had general, long term interests had been consistently and success-fully followed by the General Electric Company in the United States[1] and was afterwards to be used by a number of other major firms in that country. If Freeth's intention is construed in this way, the inter-pretation of his phrase, 'more important branches of chemistry', comes to mean those branches of chemistry which were or were likely to be of importance to the Brunner, Mond Company. This interpretation is supported by Freeth's example in the 1919 report concerning the success in the manufacture of ammonium nitrate by the skilful application of the principles of phase equilibria when the empiricists of the time had failed in attempts to operate a similar process. Of all the fields of general, long term interest to Brunner, Mond, few could have been of more obvious importance than physical chemistry. Within this somewhat ill-defined area, expertise already existed in heterogeneous equilibria in aqueous-salt systems, and the extension into the area of gas-liquid equilibria was in no sense a major alteration of course. Moreover, the potential requirements of Billingham, then so recently acquired, pointed in this direction. In brief, there was more than one vector which pointed to the route which Freeth pursued.

Hunter[2] has given the essence of Freeth's philosophy in the follow-ing words:

Freeth's experiences during the war years, when the need for scientific advice in a multitude of new problems was so desperate, led him to a clear view of what Brunner Mond's post-war research policy should be. He was convinced that it had been a source of weakness for most chemical indus-tries in Britain to depend so heavily on the universities for advice in the application of new knowledge to novel manufacturing processes. To bridge the gap between university laboratory and industrial plant, he wished to collect a senior research staff who would undertake both background re-search for the Brunner Mond companies, leaving large-scale process work to be undertaken by the works themselves, and also act as knowledgeable

---

[1] See, for example, C. G. Suits and H. E. Way (Eds), *The Collected Works of Irving Langmuir*, vol. 12, Pergamon, London, 1962.
[2] E. Hunter, *Chemistry and Industry*, 1961, pp. 2106–12.

consultants over an ever-widening range of physico-chemical problems. Furthermore, he intended that they should know, by direct contact, where to go in the universities to seek still deeper knowledge.

Freeth's transfer to London in 1927 as joint research manager of I.C.I. was, in the event, probably more useful to the Winnington programme than if he had remained in Cheshire. Amalgamation of four companies, each large and each with its own rather different traditions and outlook, is unlikely to have resulted in immediately smooth working at all levels; it is evident, for example, that there was a period of uncertainty at the end of 1926, with the appointment of a research manager over Freeth at Winnington and the proposal to discontinue the work for which Gibson had returned to England to join Brunner, Mond. In the long run, the formation of I.C.I. and Freeth's occupancy of a research management position at the centre enabled the Winnington programme to be pursued against a broader background than might otherwise have been the case.

The organization of I.C.I. as a whole and of the research departments was, however, on the basis of Divisional responsibility with central coordination. Costs of research were charged by and to the Division doing the work, irrespective of whether the subjects lay within existing interests or quite outside them. The role of Freeth and Rintoul in London was to consider whether suggested programmes would lead to duplication of effort or involved issues of company policy outside the knowledge of individual divisions. The several research departments thus preserved a good deal of independence and retained characteristic features which were still clearly recognizable in the 1950's. The research department at Winnington was the focus of physical chemistry in the whole company and, because of the Division's strong position technically and commercially, it was in a favourable state to launch new research ventures outside its traditional interests.

The relationships between Winnington, Leiden and Amsterdam, the earlier relationships in European physics and physical chemistry and the participation of the other relevant sections of I.C.I. in the background to the discovery of polythene are illustrated in Appendix II. From this it is clear how, through Donnan and Freeth in particular, the association with current chemical thought in Europe, notably in Holland and Germany, was established. In this way, Winnington renewed to some degree the earlier vital association which Mond had promoted on the industrial side with Germany, Holland and Belgium.

The second point which emerges is the reinforcement and extension of this relationship through Donnan and Gibson, as distinct from the Verschoyle and Swallow channels initiated by Freeth. Gibson's role here is specially important in that it was through him that the centre of interest in Holland shifted from Leiden to Michels in Amsterdam. Here the element of personal relationships which has been so potent a force in science, discovery and invention played its full part. Nor is the personality of Michels to be discounted in this regard, since there is no doubt that his contagious enthusiasm considerably influenced Freeth, Rintoul, the senior Billingham staff and, at a somewhat later stage, the Dyestuffs Group Research Committee. From his central position, Freeth was able to detach Gibson to work with Michels and a little later to add Perrin in a similar role. Here again, the second reinforcing string which we have noted before is again in evidence.

These central decisions, though perhaps this was not deliberately intended, activated Cocksedge and Swallow to identify themselves and Swallow's group with what was to be the main stream of thought. All these events, actual and latent, cannot but have been helpful in sustaining the high pressure research programme through the economic problems of 1930. Widespread identification in depth, developed communications, heavy commitments in training and facilities and the absence of any obvious requirement of major capital expenditure in the immediate future in the event of a successful outcome appear to have played an important part in sustaining the work, perhaps as much as any devotion to a policy of detached work. Hunter[1] has, however, made the point that though the 1930 reappraisal led to the cessation or relocation of researches with no direct connection with the established interests of the Division, it was a deliberate policy decision to continue and expand work on the effects of extreme conditions, high pressures, high temperatures and high vacua.

One interesting speculation concerns the special relationship which developed between Michels and I.C.I. The latter's actions in sending both Gibson and Perrin to Amsterdam, its substantial donation to the Van der Waals' Fund to support Michels' work and the subsequent dependence on Michels for equipment and equipment designs pose the question of whether there had not been a case for recruiting Michels as a full-time member of the staff of I.C.I. At the time, his position in Amsterdam was tenuous in the extreme, though

[1]Op. cit., pp. 2106–12.

it was soon after regularized, and on the face of it there must have been some incentive to bring him wholly into the I.C.I. orbit.

It is, however, doubtful if this possibility was seriously entertained for a number of reasons. Michels was primarily a physicist and his interests were considerably wider than those of Winnington or of I.C.I. Had he been recruited, it would probably have been difficult to give him as much support in men and facilities as he would have demanded. He had an unbounded flow of ideas which he freely communicated and, given appropriate support, he was judged to be of greater value as a consultant, a post which he held for more than twenty years.

Although the programme of work on chemical reactions at high pressures was sponsored by the Dyestuffs Group Research Committee, there appears to have been some lukewarmness about it. It did not, in fact, continue for very long having regard to its speculative character; it was subject to review within its short period of currency and seems to have been based almost entirely on ideas generated by Robinson, important though these were in giving direction to this otherwise *ad hoc* programme. The committee was in its membership a mixed one with Freeth and Rintoul, the three consulting professors and senior people from Dyestuffs Group. It may appear, though this cannot be confirmed, that, both in the case of the pressure freezing work and in the programme on high pressure reactions, there may in the first instance have been some dragooning of the Dyestuffs Group. Certainly, Swallow promoted the idea of pressure freezing work and the early stimulus for the high pressure programme may have come predominantly from Freeth, Rintoul and Robinson. Later, Thorpe suggested an embarrassing number of *ad hoc* experiments. One is here reminded of the comment of C.E.K. Mees[1] of Eastman Kodak on the conduct of research under committees:

The best person to decide what research shall be done is the man who is doing the research. The next best is the head of the department. After that you leave the field of best persons and meet increasingly worse groups. The first of these is the research director, who is probably wrong more than half the time. Then comes a committee which is wrong most of the time. Finally there is a committee of company vice-presidents which is wrong all the time.

Without taking Mees' comment too literally, it is evident that the

[1]C.E.K. Mees quoted by J. Jewkes, D. Sawers, and R. Stillerman, *The Sources of Invention*, Macmillan, London, 1957, p. 138.

Dyestuffs Group Research Committee sponsoring these high pressure programmes suffered from an undesirable degree of remoteness from those carrying out the work. In the end, Winnington found it difficult to stem the flow of suggestions when it wanted to choose its own way, which did not include a large number of *ad hoc* experiments.

It has been suggested that when Gibson and Fawcett first produced polythene in March 1933, the discovery virtually passed unnoticed. A number of reasons has been suggested for this, but perhaps the salient ones were the smallness of the yield, the current trend of thought in organic chemistry at that time and that the implied criterion for success in the programme of work sponsored by the Dyestuffs Group Research Committee would have been the production of new organic chemicals of a conventional kind. In spite of the high quality in the personnel of the committee, the common problems and limitations in carrying out research in industry under committee supervision are none the less in evidence. Where this kind of direction and control is exercised, it is almost inevitable that the criteria for success are delineated in advance actually or by implication. As a result, there is often little stimulus for the free play of serendipity—the art of profiting from the unexpected.

After the withdrawal of Dyestuffs support, a new set of forces became more significant operationally. These were associated with the challenge posed by an advanced and novel technique. It may be recalled that technique in its own right had a place in the scheme of things arising in the first place from the teaching of Kamerlingh Onnes; the translation of this to Winnington was heavily reinforced by the close association of many of the Winnington staff with Michels. In the period between the first discovery by Gibson and Fawcett and the rediscovery by Perrin and his colleagues, the considerable investment in technique development came to fruition and there was created a situation in which the high pressure laboratory at Winnington was probably unique anywhere in the world.

One of the formulae for potential success in scientific research which has come into prominence in the past ten years or so in fields as diverse as chemical physics, biochemistry, structural engineering and radio-astronomy is that there be brought together in stimulating circumstances two resources—hardware of the most advanced and up-to-date kind and better than average research staff. Though some may bewail the passing of the Rutherford era of sealing wax and string, conveniently forgetting that Cockcroft, for example, was primarily an electrical engineer, there are too many examples to make it realistic to deny that the formula has considerable validity. By 1935, this was in fact the situation at Winnington, the replacement of Gibson and Fawcett by Perrin and his collaborators possibly providing the increased stimulus that the situation demanded for productive exploitation. Fawcett's transfer to other work after the 1933 discovery is readily explicable, but Gibson has been naturally reticent about the detailed circumstances concerning his replacement by Perrin. He has commented[1] perhaps a little ruefully: 'Two months later it was obvious that the decision to put Perrin in charge of the work had been more than justified.'

The circumstances which existed in 1933 had changed significantly by December 1935, the quantity of product obtained was larger and the change in personnel were both factors likely to sharpen recognition of a positive achievement. Moreover, the programme had after many vicissitudes and abortive studies reached a stage when it was being given possibly its last major fling, and though the result was somewhat unexpected, there were this time few reservations that the discovery was significant.

If, as is suggested above, the climate of opinion had become more optimistic, the decision to build the first commercial plant on a scale of a few hundred tons a year becomes rather more comprehensible. Viewed in isolation, the decision to build such a plant using a quite novel technology, which had, however, been proved on a continuous operational basis, to produce for so limited a market has a strong entrepreneurial flavour. The prospect of adequate costing by anything approaching modern standards for well tried unit operations in chemical engineering was remote in the extreme. The intended market was a specialized one in which technological developments in high frequency transmission had created an unfilled need. If nothing other than polythene could do the job adequately, the cost of the insulant to the cable maker may have been less important in the special circumstances which apply to the cable laying industry. The insulant is but one of the component material costs and the laying of deep sea cables even in those days cannot have been a cheap undertaking in the utilization of ships and men. The price of the product initially set proved to be below the cost of production and remained so for a considerable period, but this situation was complicated by the incidence of wartime demands. In retrospect, the makers

[1] R.O. Gibson, *The Discovery of Polythene*, op. cit., p. 23.

appear to have been fortunate in being able to start with a price as high as five shillings per pound, having regard to the value of money in 1939.

There was, however, another possible factor of significance in the decision to proceed with manufacture. Some of the worst aspects of the economic troubles of the thirties were receding, and the rationalization of the chemical industry in Britain following the I.C.I. merger had largely guaranteed the monopoly position in sodium carbonate enjoyed by the Alkali Division. Thanks to the arrangement with Solvay negotiated some half century earlier and the continuous effort in process research and development at Winnington initiated by Mond and steadily continued after his departure, the ammonia-soda process had reached a state of technical maturity and had achieved some stability in relation to the caustic soda-chlorine limb of the alkali industry. The outlook was, in consequence, secure and the case in general terms for the employment of an element of risk capital in a new venture was not wholly unfavourable.

A further point was that the Wallerscote plant was in many respects a scaled up version of the experimental facilities, and because of its unusual technology the contribution of the scientists as distinct from the engineers to its design was greater than would be normal. As a result, the handing over of the process to the developers in any formal sense was necessarily blurred. Where there exist rather specific units in an organization formally charged with defined responsibility for a new project as it moves from the research laboratory into the hands of production teams, there tends to arise, by direction or as an inherent operating characteristic, and motivated amongst other things by self preservation, a series of checks and reassessments at the points of transfer. The first commercial plant at Wallerscote by its very nature probably escaped much of this re-evaluation.

It is of some interest, though not particularly fruitful, to speculate on what might have happened in the development of polythene had the Second World War not intervened. It seems probable that the emergence of polythene from a polymer for a specialized use to one suitable for a wide variety of applications would have taken place more rapidly, if we are to judge from the actions of Union Carbide once the wartime restrictions were removed. The geographical centre for product development is likely to have been the United States with or without the war because of the existence there of more experience, and often more competitive vigour in product marketing, and fewer inhibitions in the engineering industry on which large scale production and utilization in a variety of forms heavily depended.

The United States which, even at that time, had a substantial petrochemical industry was in a specially favourable position potentially to produce ethylene in quantity at a cost which European producers would have found hard to match. The ethylene originally used by du Pont was obtained by fractionating coke oven gas; Union Carbide used ethylene from their established process involving the extraction of ethane and propane from natural gas; ethylene from oil was still in an embryonic state and was largely stimulated by the growth in demand for polythene in the post-war years.

While licensing arrangements may have been contrived rather differently in the absence of wartime requirements, perhaps the major disadvantage suffered by I.C.I. was the fact that, having been the inventors, they found themselves after the war in a retarded position with regard to their own production, a situation which was not improved by the anti-trust judgment. In maintaining the original price which was below the early costs of production, they had also accumulated some losses.

The invention of processes for making high density polythene other than by the use of high pressures, while differing in background and detail from the Winnington work, contain some elements in common with it. The Ziegler process and those of Standard Oil of Indiana and Phillips Petroleum arose, as in the work at Winnington, as a result of programmes of research being pursued on a long-range basis. The origins of Ziegler's interests in metal-alkyls in general have been described in some detail and the efforts of the petroleum companies to convert product streams of low commercial significance to ones of higher value have been and continue to be an important activity in the petroleum industry. Over the years, this has yielded a number of innovations[1] of the cracking and reforming type. The existence of polythene, even with somewhat different properties as a market product, largely ensured the recognition of these new discoveries in their very early stages. It appears, however, that the Ziegler process has enjoyed more favourable development by Hoechst than the others, and the reasons for this may in general be associated more with the ultimate potential demonstrated by Natta's work, than with

[1]See, for example, J. L. Enos, 'Invention and Innovation in the Petroleum Refinery Industry' in *The Rate and Direction of Inventive Activity*, Princeton University Press, Princeton, New Jersey, 1962, p. 299.

the fact that the process avoids the use of high pressures and yields material of higher density and crystallinity.

Many of the early would-be Ziegler licencees before Natta's contribution were, however, probably seeking a route free from the I.C.I. patents, either because they wished to be free, or could not get the know-how as well as the patents. This has particular relevance to Germany, where Badische Anilin Soda Fabrik had the monopoly of the I.C.I. patent, though they did not use the I.C.I. process. Hoechst were, therefore, in an admirable position to develop a new thermoplastic.

The question of substitutability of Ziegler high density polythene for low or medium density polymer made by the I.C.I. process is beset in some parts of the world by tariff considerations and is not related exclusively to matters of preferred applicability or use. The Ziegler product is often basically more expensive and, in the absence of complicating factors, makes its way in certain uses for which the high pressure polymer is unsuitable.

A feature of the discovery of the Ziegler catalysts was the part played by the increased knowledge in polymer science generated in the period between the I.C.I. work and that of Ziegler. The X-ray and infra-red studies of Bunn and others clarified many of the concepts of crystallinity and linearity in polymers, and a good deal of systematic knowledge on the kinetics of polymerization reactions, especially on the role of inhibitors, had accumulated. One of the early books on this subject was that of C. E. H. Bawn[1] published in 1948. In addition, there was after 1945 a strong revival in inorganic chemistry, initiated in part by the wartime work on nuclear weapons, which later generated a great deal of research on metal-organic compounds including both coordination compounds and those in which the metal atom was directly bonded to a carbon atom as in the metal-alkyl compounds. Ziegler's extension of his long established interest in the alkali metal-alkyls to other metallic elements such as aluminium was, in the light of the resurgence of interest in this field, less fortuitous than it may superficially appear. Furthermore, having suspected traces of nickel as being inhibitors in the propagation of short polymer chains produced with aluminium alkyls alone and having decided to test systematically the effects of other possible metals, the choice of the transition metals, particularly of the first series, was a quite natural one.

[1] C.E.H. Bawn, *The Chemistry of High Polymers*, Butterworth, London, 1948.

Without wishing to press this thesis too far, it is well to recognize that the Ziegler catalysts arose out of a multifaceted background of organic and inorganic chemistry and, significantly, in the less well defined area which lies between these two traditional fields. In brief, there is some justification in the claim that this discovery arose by the synthesis in the mind of one man of particular developments in the field of chemistry in which the vital operation was the preparation of new compounds—in this case, a catalyst of unusual type and properties. In partial contrast, the I.C.I. discovery emerged more from a background of advanced physical technique and quantitative measurement in deliberately pursuing the effects of high pressures on chemical reactivity. The latter depended on devising special pieces of equipment and making them work, and was, in this respect, the natural child of the ammonia-soda process.

The circumstances, the timing, the nature and location of the I.C.I. discovery all contributed to a situation in which little if any background research was undertaken by this company on alternative methods of making polythene. Some excellent work was done on compounds related to the Ziegler catalysts, but this was located in another wholly detached centre having no formal relationship with the divisions of the company and viewed with some suspicion by most of them. Moreover, much of the I.C.I. research effort on polythene in the years immediately following the war was, quite reasonably, devoted to improving the high pressure process from the point of view of increasing the scale, achieving better control and consistency of product and of elucidating the physical and mechanical properties with respect to an increasing range of possible applications. Production and process research and development were firmly rooted in Alkali Division and the pattern followed was one which to this Division with its long background of evolutionary development of the ammonia-soda process was second nature. Had this work been located in a Division accustomed to the risk of having its products and processes rapidly outmoded by deliberate competitive research, a watching brief on alternative developments might have been a normal precaution. This was not the case and the Ziegler-Natta developments undoubtedly took I.C.I. completely by surprise.

While these events clearly point to the now widely recognized need once a major discovery has been made for the inventors to explore actively a wide range of peripheral areas and activities which could lead to competitive developments, they also emphasize the problems confronting management in respect of the organization and control of major new ventures. In any large divisionally structured organization, there is properly some competitive spirit between divisions which is capable of profitable exploitation. To have withdrawn the polythene project wholly from the Alkali Division in the years immediately following the war would have created a considerable disturbance and it is possible that the overall effect would have been detrimental. In the event, the process side was left with Alkali Division while product development came increasingly within the ambit of Plastics Division. Hindsight may suggest alternative approaches, but these have always to be evaluated against the historical background and all the detailed circumstances of those far from normal years.

The inventive acts that have been considered arose from scientific research and on this ground qualify as scientific inventions as distinct from craft inventions. Any synthesis of the experience which this history provides is therefore restricted to situations of this kind and to the innovations which flow from them. A dominant feature of the body of knowledge we call science is its continuity. All scientific research, whatever its motive or excuse, aims at contributing to a greater or less degree new knowledge to the already accumulated store; each contribution rests upon and may in large measure be related to what has gone before and each serves to generate or point to future paths of endeavour. A second characteristic of this continuously growing body is its organization broadly in two kinds of framework. One of these is simply a framework of convenience, an arbitrary division into disciplines—physics, chemistry, geology and so on—and into hybrid and finer subdivisions, for example, biochemistry, geophysics, inorganic chemistry, microbiology. Such an arbitrary and temporary framework is of no fundamental significance.

The second type of framework in the organization of science stems from the several progressive levels of certainty, from its generalizations, its laws, its theories and its general theories. While few men and women in science seem capable of organizing or ordering knowledge on a grand scale, or even on a more modest scale, all seek in one way or another to create ordered domains within the knowledge continuum. Some of these domains are very small, others are more extensive; some represent perhaps little more than a linear correlation between two observables while others are multidimensional. As in solid substances generally, multidimensional arrays are seldom perfectly ordered. For one thing, they necessarily have boundary discontinuities, they may be internally imperfect with cracks, vacancies, pseudomorphic growths, dislocations and so on. The ordering may be so imperfect as to approximate to a liquid or an amorphous solid. If, however, the degree of order is sufficiently great to distinguish it from the continuum in which it has been created, we may in the limit think of it as the crystallization of a new phase.

By way of illustration, let us consider first the simple physico-chemical system of a solution of salt in water. At a given temperature and pressure there is a fixed amount of salt which will remain permanently in a given amount of water at equilibrium. This is a saturated solution. It is possible, however, under certain circumstances to prepare supersaturated solutions containing more dissolved salt than in the equilibrium saturated solution. Such solutions are said to be metastable and may continue indefinitely in this state in the absence of nuclei on which the surplus salt may crystallize out and restore the supersaturated solution to the equilibrium state. In the continuum of the largely disordered supersaturated solution the nucleus for a new, ordered (solid) phase represents a degree of order on a very small scale. Nuclei may form spontaneously when the solution exceeds infinitesimally the saturation concentration, or they may form, and then only with difficulty, when a comparatively high degree of supersaturation exists. Once formed, they may grow rapidly or they may redisperse; many may be simultaneously created and grow at comparable rates; many may be formed initially, but only a few grow and the remainder redisperse or die as growing entities. The resulting solid phase may be highly ordered over large regions, for example, in the formation of large single crystals, or the order may extend over many small individual volumes as in fine polycrystalline precipitates. Moreover, the rate at which the change from metastability to the equilibrium state takes place will be governed, among other things, both by the rate of nucleus formation and by the rate of nucleus growth.

Contributions to the grand sum of scientific knowledge in their ultimate unitary form of experimentally verifiable facts each add to the degree of supersaturation of the solution—the knowledge continuum. Depending on the degree of supersaturation, nuclei representing small regions of latent order may form, if conditions are favourable. There may be many or few nuclei depending on the size, structure and the peculiarity of the way in which the components have to be put together. A whole series of incipient nuclei may form almost simultaneously by the addition of some key fact or set of facts that was hitherto missing, or a small number of larger nuclei of a high order may form under these circumstances. As in the physico-chemical analogy, nuclei may grow thereby enhancing their stability and extending the domain of order in a particular area of knowledge, or they may disperse, or they may coalesce. They may grow at the same or at different rates, perhaps the larger at the expense of the smaller. The variety of possibilities is as large as there are people in whose minds the synthesis occurs, and the irreproducibility of be-haviour almost as unpredictable.

Contributions to knowledge in science take many forms. Those worthy of written record, papers, reviews, books, patents and general articles seek not only to convey units of knowledge, but also usually to attempt some degree of ordering however small. The result may be as varied as the nuclei, few many, stable or unstable, with high or low growth potential to yield small or large regions with great variety in the degree of order. Every patent or other record of an invention represents a packet of knowledge, a nucleus, born in the continuum of knowledge and necessarily related to it. Their dis-tinction from other kinds of packets lies more in the future than in the past, since effective recognition in practical terms is accorded an invention if it ultimately leads to innovation, thereby supplying an existing or potential material need or service. Its future as a nucleus is conditioned not only by the knowledge continuum of its birth, but also by its external surroundings with which it interacts. But inter-action with an external environment both before and after the nucleation step may cause many changes in behaviour. As in our supersaturated salt solution, raising the temperature may, for ex-ample, decrease the degree of supersaturation, lowering the tempera-ture enhance it. The introduction of extraneous particles may cause premature nucleation; scratching the container promote it; inhibitors to crystallization or agents capable of modifying the number, type, orientation and morphology of the nuclei and the ultimate solid phase may be introduced. In some circumstances, all or some of these possi-bilities may be helpful, while in others they may be diversionary or may cause unwanted results.

In the discovery and early development of polythene in Great Britain and in the later work in Germany and the United States, many of the features noted above are already evident—the different areas of the knowledge continuum which gave them birth, the differing nature and rate of the processes of nucleus formation and growth, the differences in the character and extent of interaction with the external environment, the varying personalities and backgrounds of the human catalysts.

III

TERYLENE

Introduction

Early in 1941, J. R. Whinfield and J. T. Dickson, working in a small laboratory of The Calico Printers' Association Limited at Accrington near Manchester, prepared the first sample of a high molecular weight, fibre-forming, synthetic polyester of terephthalic acid and ethylene glycol, polyethylene terephthalate, which today is known as Terylene in the United Kingdom and Dacron in the United States. At the time, Whinfield was nearing forty years of age, while Dickson was in his early twenties, and though the initial experiment was actually carried out by Dickson, it was in Whinfield's mind that the ideas underlying it had crystallized. These were derived from the sources outlined below.

The first was in the nature of Whinfield's early career which, after leaving Cambridge, had included a short period with the consulting firm of Cross and Bevan, whose principals had invented viscose silk in 1892 and an industrial process for the manufacture of cellulose acetate in 1894. A second and most significant element stemmed from the work of W. H. Carothers on polymerization carried out at the Experimental Station of the du Pont Company at Wilmington, Delaware in the period, 1928–37. A third component arose from the particular conditions which existed in the relevant period in The Calico Printers' Association Limited, and to the general scientific and technological climate surrounding polymers and polymerization during the 1930's. In the sections which follow, we shall discuss each of these contributions in turn and try to assess its influence on the ultimate discovery.

The preparation in a laboratory of a few grams of a potentially interesting new material is but the first step in an innovation. The transformation of such a discovery first into an effective patent, then through subsequent steps of development and ultimately to the construction of plant and production on an industrial scale involve the solution of many problems by many hands in a co-ordinated manner. The progress from invention to innovation is seldom a simple, direct process and there are comparatively few examples where the relevant issues at each stage can be delineated. In the case of Terylene, these steps can be described in some detail and afford an insight into some of the problems posed by this transformation.

Today, Terylene (and Dacron) is a synthetic fibre of international significance. The first full scale plant in the United Kingdom, originally designed with a capacity of 11 million pounds per year, came into operation early in 1955. Currently, modern, individual plants in the United Kingdom, such as that at Kilroot near Carrickfergus in Northern Ireland, often approach 20 million pounds per year and this figure is expected to rise to 40 to 50 million pounds by 1968.[1] In the United States, the estimated total plant capacity for this polymer is expected to be 426 million pounds by the end of 1965, of which du Pont controls some 240 million pounds per year. By way of comparison, the installed capacity for Nylon in the United States at the same date is given as 1,075 million pounds per year.[2]

Polymers of industrial and commercial importance with molecular weights in excess of about 10,000 may be grouped in three broad classes:

(i) natural polymers, for example, silk, wool, cotton, natural rubber.

(ii) derived polymers obtained by physical or chemical modification or reconstitution of natural polymers, for example, nitrocellulose, cellulose acetate.

(iii) synthetic polymers in which the high molecular weight substance is obtained by a polymerization reaction from one or more simple monomeric substances of low molecular weight of the order of a hundred.

Historically, the derived polymers of cellulose constitute the beginnings of industrial polymer chemistry dating from the invention of celluloid in 1862, and though this class continues to contribute significantly to industrial polymers as a whole, there are inherent limitations in the potential range of products and properties obtainable where one of the starting materials is a natural polymer. Such a limitation is absent in the area of wholly synthetic materials. It is, however, significant that many of the ideas about polymers and some of the techniques for studying their properties and elucidating their structures were developed at a time when natural and derived polymers commanded nearly the whole scene.

[1] *The Guardian*, 22 October 1965.
[2] *Chemical and Engineering News*, *43*, no. 36, 24 (1965).

Synthetic polymers are conveniently sub-divided into two classes:

(i) condensation polymers formed by multiple condensation reactions between chemically reactive functional groups attached to
one or more species of reactant. Polyamides of the Nylon type, polyesters of which Terylene is an example, and a wide range of alkyd
resins, urea-formaldehyde and phenol-formaldehyde resins are typical
members of this group.

(ii) addition polymers formed by repeated addition reactions between chemically reactive monomers, for example, the vinyl polymers
based on monomers of the type, $CH_2{=}CHX$, where X may be a
hydrogen atom, a halogen atom, an acetate, phenyl, cyanide or other
group capable of conferring the necessary chemical reactivity on the
molecule.

Another method of classifying polymers is related to their ability
to form strong, flexible fibres. Some are capable of being spun or
drawn from a melt or solution and are the potential candidates
for development as synthetic fibres for use, for example, in the manufacture of textiles. Others do not display this property either at all or
to a sufficient degree, or, if they do so, yield fibres which lack strength
or flexibility. The latter products are likely to find uses in the manufacture of moulded articles, films and coatings. Those falling in the
first category may also often be utilized in the second. This difference
in properties was one of the interests of the polymer scientists and
technologists of the 1920's and 1930's who were anxious to determine
the relationship between polymer properties and structure on a
molecular scale. If these relationships could be satisfactorily eluci-
dated, there would be good prospects of predicting the kinds of
monomers which would be worth studying as a means of producing
useful, new polymeric materials for one or other purposes.

Whinfield at Cross and Bevan's

Whinfield graduated from Gonville and Caius College, Cambridge,
in 1922 at a time when Cambridge chemistry as a whole was dominated
by W. J. Pope, who occupied the chair from 1908 until his death in
1939. In Whinfield's time, the two members of staff who had most
influence on the Part II students were W. H. Mills and E. K. Rideal.
Pope's interests and his research achievements in chemistry were considerable.[1] The dominant theme was stereochemistry, and neither he

[1] *Royal Society Obituary Notices of Fellows, Sir William Jackson Pope, 3,*
291–324 (1939–41).

5

nor his collaborators ventured into such complex technical fields as cellulose chemistry, then very much the undisputed province of the industrial chemist. Whinfield did not remain for postgraduate research and, as a result, Cambridge appears to have contributed little in influencing him towards the field of chemistry in which he was to spend the greater part of his professional life. The same cannot be said of his first postgraduate year with Cross and Bevan.

C. F. Cross (1855–1935) was educated at Kings College, London, Zürich University and Polytechnicum and at Owens College, Manchester, before taking the B.Sc. degree of London University in 1878. At the instigation of Roscoe of Owens College, he then immediately entered the field of cellulose chemistry, undertaking research for the Barrow Flax and Jute Company on the constitution of jute fibres. This work was continued at the Jodrell Laboratories at Kew, where he became associated with E. J. Bevan, a former fellow student at Manchester, who had spent some years in the paper industry. This association led Cross and Bevan in 1885 to establish a practice as analytical and consulting chemists at New Court, Lincoln's Inn, London, which Whinfield joined as a 'pupil' after leaving Cambridge.

Bevan died in 1921, and in Whinfield's time Cross was a comparatively old man of sixty-eight. The work of the laboratory was then the usual, varied mixture characteristic of the public analyst and consultant of the day, but in the preceding years until about 1920, Cross and Bevan had established international reputations in cellulose chemistry and technology. E. F. Armstrong, in the Royal Society obituary notice of Cross, who was elected a Fellow in 1917, aptly wrote,[1] 'After all, for many years Cross was cellulose.'

Cross's reputation was based on many achievements of which probably the greatest was his discovery in 1892 with Bevan and Clayton Beadle of a method of obtaining cellulose in a soluble form. By treating cellulose, first with aqueous caustic soda and then with carbon disulphide, a golden yellow viscous liquid of cellulose sodium xanthate was obtained. This liquid, when injected in a fine stream into an acid bath, yielded fibres which, after treatment to remove the sulphur, consisted of pure, regenerated cellulose. This viscose process for which the patent ran its full term of years without serious challenge was not developed and used commercially without some difficulties, but by 1902, when it was taken up energetically by Courtaulds, the future of the viscose rayon industry was largely

[1] *Royal Society Obituary Notices of Fellows*, *1*, 459 (1932–5).

assured. To many of the companies in Europe engaged in working the process, Cross acted as a consultant.

The second achievement on which Cross and Bevan's reputation was built concerned another modified cellulose polymer, cellulose acetate. In 1894, Cross and Bevan took out the first industrial patent for its manufacture, and in the years which followed, they were heavily engaged in consulting work related to this process. Another interest centred on nitrocellulose, the first of the derived cellulose polymers. In 1901, Cross, Bevan and Jenks showed that the primary cause of instability in this material which had caused much trouble was due to the presence of sulphuric ester residues remaining from the nitration of the cellulose with a sulphuric-nitric acid mixture.

The third area of endeavour of Cross and Bevan was in paper making and, while in this there were fewer signal advances, their overall contribution, especially in respect of paper quality, was considerable. Their reputation in this field was effectively established by the publication in 1888 of their book on paper making which by the 1930's had run to five editions.

Cross and Bevan's widespread interests in cellulose chemistry and technology are well illustrated in their many published papers in this field which appeared during the period, 1880–1920, in their classic textbook[1] and in a series of volumes on researches on cellulose,[2] of which the first appeared in 1895 and the fourth in 1922. Of the textbook, Armstrong[3] wrote in 1935:

The book is unique; full of imagination and inspiration, with ideas often partly expressed tumbling over one another; it is a memorial of the true Cross. It is not easy today to assess the value of his contributions to this subject. Cellulose has been the most elusive of substances, and though there are signs that at long last we are on the track of its make up, the knowledge is still only in a tentative state. The modern work has taken little heed of Cross and his theories of structure, and full recognition is lacking of his appreciation of its colloidal behaviour as an amphoteric electrolyte.

From this brief description we may derive some ideas on the possible impact which Whinfield's sojourn with Cross and Bevan had upon him. In this period, he was brought face to face with the hub of

[1] Cross and Bevan (C. F. Cross, E. J. Bevan and C. Beadle), *Cellulose*, Longmans Green, London, 1895.
[2] Cross and Bevan, *Researches on Cellulose*, vol. I, 1895–1900, (1901); vol. II, 1900–1905, (1906); vol. III, 1905–1910, (1912); vol. IV, 1910–1921, (1922), Longmans Green, London.
[3] *Royal Society Obituary Notices of Fellows*, *1*, 459 (1932–5).

cellulose chemistry in Britain and possibly in Europe at that time. Because of the lack of definition in many aspects of this field, it is probable that Whinfield acquired some speculative traits with ultimate practical use perhaps not too far in the background. Moreover, the possible relationships between properties and structure and the prospects of a more rational approach to problems in this field were not wholly swamped by purely technological needs.

Cross has been described as a man of striking appearance, full of culture, with broad interests and widely popular, and one to whom science and the pursuit of knowledge for its own sake is said to have meant as much as the technological developments of the cellulose industry. In a sense, he was a knowledge entrepreneur. This, then, was the man and this the climate in which Whinfield spent his earliest postgraduate years. He has publicly acknowledged[1] Cross's influence upon him and his future, and it was, in fact, Cross who secured his original appointment to The Calico Printers' Association Limited. Whinfield has also recalled that at the time of this appointment, the Shirley Institute was just beginning its basic work on cellulose in which he took an immense and sustained interest.

The work and influence of W. H. Carothers

The short, brilliant and tragic life of W. H. Carothers of forty-one years, which terminated on 29 April 1937, falls into three periods, his early years, his chemical education and early researches, and the final period beginning in 1928 with the du Pont Company. Our concern is with the last two periods. Carothers graduated in 1920 with the degree of Bachelor of Science in chemistry from Tarkio College, Tarkio, Missouri and completed the requirements for the degree of Master of Arts at the University of Illinois in the summer of 1921. During the academic year of 1921–2 he was an instructor in analytical and physical chemistry at the University of South Dakota and simultaneously began his first research. Characteristically, this was based on an idea concerning valence electrons originally advanced by Langmuir in 1916. The aim was to examine the implications of this idea in relation to organic chemistry and in his second, independent paper[2] Carothers gave one of the first, definitive, workable applica-

---

[1] J. R. Whinfield, *Nature*, *158*, 930 (1946).
[2] W. H. Carothers, *Journal of the American Chemical Society*, *46*, 2226–36 (1924).

tions of the electronic theory to organic chemistry.

Carothers returned to the University of Illinois in 1922 and completed his work for the Ph.D. degree in 1924 under the supervision of Dr Roger Adams. Though his major preoccupation was organic chemistry, his skills in physical chemistry were also widely recognized at this time. He remained as an instructor at Illinois for a further two years, before moving to an instructorship in organic chemistry at Harvard in 1926. Of his two years at Harvard and an appreciation of his research capabilities, J. B. Conant, then Professor of Organic Chemistry, wrote in these terms:[1]

In his research, Carothers showed even at this time that high degree of originality which marked his later work. He was never content to follow the beaten track or to accept the usual interpretation of organic reactions.

The papers published on his research work of this period contain no specific reference to polymers or to polymerization, but through them runs the strongly developed theme of the elucidation and interpretation of chemical reactivity in organic reactions. This formed the starting point of his later work, and the recognition of its possible relevance to polymerization reactions belongs, according to Conant, to his Harvard days.

In 1928, the du Pont Company planned to embark on a new programme of fundamental research at its central laboratories, the Experimental Station at Wilmington, Delaware and chose Carothers to head the work in organic chemistry. The principal attractions of the post were the supply of trained assistants, the opportunity to pursue research without teaching responsibilities and the freedom to work on problems of his own choice. In the nine years that followed, Carothers made several major contributions to the theory of organic chemistry and at the same time produced discoveries which led directly to new materials of great commercial importance. If anything, the former rather than the latter, at least in the earlier years, seem to have been the main driving force.

Carothers' papers in his years at Wilmington, almost all of which were published in the *Journal of the American Chemical Society*, are in the form of two major series. One group in eighteen parts under the collective heading of 'Acetylene Derivatives and Synthetic Rubber' appeared between 1931 and 1934. The second series of twenty-nine

[1] Roger Adams, *National Academy of Sciences of the United States of America Biographical Memoirs*, *20*, 293–309 (1939).

numbered and three unnumbered papers under the general heading of 'Polymerization and Ring Formation' were published in the years 1929 to 1936. These papers, as well as those published in the earlier years, were collected and republished[1] in 1940.

The work in the first of these groups comprised an exhaustive study of acetylene polymers and their derivatives. The approach was a synthetic one and amongst the many compounds prepared was monovinylacetylene. Carothers found that this reacted with hydrogen chloride to form 2-chloro-1, 3-butadiene which he called chloroprene. This substance which was analogous to isoprene polymerized several hundred times as rapidly as the latter, giving a product which was superior in its properties to all previously known synthetic rubbers. It was, for example, the first synthetic material to reproduce the property of natural rubber of developing a fibrous orientation when stretched and reverting to a disordered, amorphous state when the stress was removed. This discovery was developed by du Pont, the polymers of chloroprene being marketed under the now well known trade name of Neoprene.

The second group of papers on polymerization and ring formation which is of special interest in the present context also constituted an important contribution both to organic chemistry and to the chemical industry. The background of this work has been preserved in a letter written by Carothers to Dr J. R. Johnson of Cornell University dated 14 February 1928. It reads, in part, as follows:[2]

One of the problems which I am going to start work on has to do with substances of high molecular weight. I want to attack the problem from the synthetic side. One part would be to synthesize compounds of high molecular weight and known constitution. It would seem quite possible to beat Fischer's record of 4,200. It would be a satisfaction to do this, and facilities will soon be available here for studying such substances with the newest and most powerful tools.

Another phase of the problem will be to study the action of substances xAx on yBy where A and B are divalent radicals and x and y are functional groups capable of reacting with each other. Where A and B are quite short, such reactions lead to simple rings of which many have been synthesized by this method. Where they are long, formation of rings is not possible. Hence reaction must result either in large rings or endless chains. It may be possible to find out which reaction occurs. In any event the reactions will

[1] Collected Papers of Wallace Hume Carothers on High Polymeric Substances, ed. H. Mark and G. S. Whitby, Interscience, New York, 1940.
[2] Roger Adams, op. cit.

lead to the formation of substances of high molecular weight and containing known linkages. For starting materials will be needed as many dibasic fatty acids as we can get, glycols, diamines, etc. If you know any new sources of compounds of these types I should be glad to hear about them.

Seldom has there been such a specific statement of research aims which were sustained throughout the series. Before describing those aspects of Carothers' work which are pertinent for our present purposes, it is necessary to comment briefly on the general state of knowledge in polymer chemistry at that time. Until the turn of the century, it had been widely believed that what we now know as polymers were impure aggregations of molecules of ordinary size, and that if sufficient purification of these resinous masses was carried out, individual crystalline compounds of the normal type should be obtained. The next stage was embodied in the so-called association theory which predicated the idea that in polymers the molecules were bound together by forces different from those in ordinary chemical bonds, and that the physical and chemical behaviour of these substances were manifestations of these special forces.

By the early 1920's, largely as a result of Staudinger's work on synthetic polyoxymethylenes, these early views had been replaced by the molecular chain theory which recognized that polymerization could lead to the formation of long chains built up by ordinary chemical bonds formed between suitable molecules assembled in a more or less regular fashion. At the same time, X-ray analysis of natural fibres such as cellulose, wool and silk had shown them to be microcrystalline and in this respect apparently different from the amorphous, resinous synthetic polymers known when Carothers began his work. There was an additional problem posed by the X-ray results, namely, that the unit cells of the crystal lattices calculated for the microcrystalline natural materials were too small to contain large molecules, it being wrongly assumed at the time that the unit cell could not contain less than one molecule.

In the first paper in the series, Carothers distinguished between addition polymers formed by recurring addition reactions between monomer molecules and condensation polymers in which the structural units in the polymer were joined as a result of multiple condensation reactions involving the elimination of a small molecule, such as water or ammonia, arising from the combination of functional groups initially existing in the monomer molecules. It is this latter group with which the body of Carothers' work of current interest is concerned. The functional groups of principal concern were the hydroxyl, $-OH$, the amine, $-NH_2$, and the carboxyl, $-COOH$, occurring bifunctionally in molecules mainly of the following types:

$$HO-R_1-OH \text{ (glycols)}$$
$$HOOC-R_2-COOH \text{ (dicarboxylic acids)}$$
$$HO-R_3-COOH \text{ (hydroxycarboxylic acids)}$$
$$H_2N-R_4-COOH \text{ (aminocarboxylic acids)}$$
$$H_2N-R_5-NH_2 \text{ (diamines)},$$

where R_1, R_2, R_3, etc. represent organic diradicals.

Carothers recognized that such bifunctional molecules might react intramolecularly, for example,

$$HO-R_3-COOH \rightarrow R_3 \begin{array}{c} \nearrow CO \\ | \\ \searrow O \end{array} + H_2O \qquad (1)$$

or intermolecularly, for example,

$$HO-R_3-COOH + HO-R_3-COOH \rightarrow$$
$$HO-R_3-CO-O-R_3-COOH + H_2O \qquad (2)$$

or

$$H_2N-R_5-NH_2 + HOOC-R_2-COOH \rightarrow$$
$$H_2N-R_5-NH-CO-R_2-COOH + H_2O \qquad (3)$$

Ring formation represented by (1) was preferred when the diradical R was such as to lead to a 5- or 6- or sometimes a 7- or 8-membered ring, but in other cases the intermolecular-type reaction was the favoured one. The repeated reactions of the types (2) or (3) should lead to the formation of long linear molecules, for example,

$$HO-R_3-CO-O-R_3-CO-O-R_3-CO-O-R_3-COOH$$

or

$$H_2N-R_5-NH-CO-R_2-CO-NH-R_5-NH-CO-R_2-COOH$$

and so on.

The essential issues studied by Carothers were how the physical and chemical properties of the high polymers depended on their molecular weight, that is the average length of the chain, the type and nature of the structural units in the chain and on the nature of the linkage. In pursuit of this aim, a large number of bifunctional-type molecules was examined both in respect of type (2), where the two functional groups on the molecule were different and mutually reactive, and in those of type (3) involving two different bifunctional molecular species. An important feature was that, virtually without exception, the work was directed to the formation of linear polymers using bifunctional reactants. The formally similar reactions between polyhydric alcohols and polybasic acids, where the functionality of at least one of the reactants was greater than 2, were excluded, it being known by 1929 that such reactions gave resinous products generally considered to be amorphous. The exclusion of reactions of this type was not, however, made primarily on these grounds, but because in such systems it was not possible to assign with any confidence a structure from a knowledge of the synthesis reaction.

Of the several systems studied in detail, two were predominant, the polyesters involving the $-CO-O-$ ester linkage and the polyamides based on the $-NH-CO-$ amide linkage. There is a number of variations possible within these types, but the details need not concern us here. In the earlier papers, emphasis was placed on linear polyesters with molecular weights from 800 to 5,000, referred to as α esters, derived from the acids of the series

$$HOOC-(CH_2)_x-COOH$$

and glycols of the series, $HO-(CH_2)_y-OH$. These polyesters were all microcrystalline solids which melted around 80°C, dissolved readily in solvents such as chloroform and ethyl acetate giving solutions which displayed colloidal behaviour, but were not highly viscous.

In the twelfth paper of the series, the preparation and characterization of superpolyesters of molecular weights in excess of 10,000 using a molecular still were described. The polyesters of molecular weight greater than 5,000, called ω esters, were materials of considerable hardness and toughness and, when placed in solvents, first imbibed the solvent and then swelled, ultimately giving highly viscous solutions. On heating to about 80°C they melted and became transparent, but flow only ensued at higher temperatures. In the solid state, these products gave sharp X-ray patterns which closely resembled the patterns for the corresponding microcrystalline α esters.

At a few degrees below the melting point, the sharp pattern began to disappear and at the melting point only the diffuse halo characteristic of the amorphous liquid appeared. In the plastic state, Carothers found that superpolyesters could be readily drawn out into strong, pliable, highly oriented fibres, this being the first discovery of the fibre-forming properties of polyesters.

The paper by J. W. Hill and Carothers describing this work was submitted for publication to the *Journal of the American Chemical Society* on 12 November 1931 and appeared on 6 April 1932. A report of the work was also given to the Buffalo meeting of the American Chemical Society in the autumn of 1931 and led *Time* magazine to observe:[1]

The Carothers-Hill fibre is as lustrous as silk, stronger and more elastic than rayon and as strong and elastic as real silk. It is too expensive to manufacture commercially; it is mainly a demonstration of chemical knowledge and skill.

This comment might have led the reader to conclude that the cost of manufacture was the only barrier to commercial development. What the press report omitted to mention was the low melting point, around 80°C, and the fact that these polyesters had poor hydrolytic stability, both features which would, in any case, rule out their widespread practical use.

In the next paper submitted and published at the same time, Hill and Carothers turned their attention to polyamides and found that, even at a molecular weight of 3,000, these polymers were tougher, less fusible and less soluble than the superpolyesters, a fact which they attributed to the higher molecular cohesion of the $-CO-NH-$ group. The first polyamide prepared was based on ε-aminocaproic acid, which gave both a seven membered cyclic lactam and a linear polymer of the form

$$-NH-(CH_2)_5-CO-NH-(CH_2)_5-CO-NH-(CH_2)_5-CO-$$

This polyamide was so infusible and insoluble, even at a molecular weight of 3,000, that its capacity to form fibres could not readily be tested. Not surprisingly, Hill and Carothers then sought to prepare mixed linear polyesters-polyamides containing both $-CO-O-$ and $-CO-NH-$ type linkages and found their properties intermediate between the corresponding polyesters and polyamides.

[1] Quoted by E. F. Izard, *Chemical and Engineering News*, 32, 3724–28 (1954).

The work of Carothers and his colleagues to this time not only on polyesters and polyamides, but also on polyanhydrides and polyacetals enabled them to publish in April 1932 in the fifteenth paper of the series a fairly complete description of the behaviour of superpolymers when they were treated mechanically, and especially when drawn as fine filaments from the melt or dry spun from a solution in chloroform. The discussion section of this paper brings together a number of factors which are specially relevant to the discovery of Terylene and is, therefore, reproduced here in full:[1]

Our studies of polymerization were first initiated at a time when a great deal of scepticism prevailed concerning the possibility of applying the usually accepted ideas of structural organic chemistry to such naturally occurring materials as cellulose; and its primary object was to synthesize giant molecules of known structure by strictly rational methods. The use of synthetic models as a means of approach to this problem had already been undertaken by Staudinger, but the products studied by him (polyoxymethylene, polystyrene, polyacrylic acid, etc.) although unquestionably simpler than naturally occurring polymers, were produced by reactions of unknown mechanism, and their behaviour, except in the case of polyoxymethylene, was not sufficiently simple to furnish unequivocal demonstration of their structure. On the other hand, the development of the principles of condensation polymerization described in preceding papers of this series has led to strictly rational methods for the synthesis of linear polymers, and the structures of the superpolymers III to VI follow directly from the methods used in their preparation. Meanwhile the weight of authoritative opinion has shifted; further evidence has accumulated and cellulose has been assigned a definite and generally accepted structure. Like the synthetic products III to VI, it falls in the class already defined as linear superpolymers.

I

Cellulose (polyacetal)

II $-NH-R-CO-NH-R'-CO-NH-R-CO-NH-R'-CO-$

Silk (polyamide)

III $-O-R-CO-O-R-CO-O-R-CO-O-R-CO-O-R-CO-$

Polyester (from hydroxy acid)

[1] J. W. Hill and W. H. Carothers, *Journal of the American Chemical Society*, 54, 1579–87 (1932).

IV $-O-R-O-CO-R'-CO-O-R-O-CO-R'-CO-$

Polyester (from dibasic acid and glycol)

V $-O-R-CO-NH-R'-CO-NH-R'-CO-O-R-CO-$

Mixed polyester-polyamide

VI ... $-O-CO-R-CO-O-CO-R-CO-O-CO-R-CO-$

Polyanhydride

Addition polymers of very high molecular weight have been synthesized in the past, but the capacity to yield permanently oriented fibers of any considerable strength has not been observed hitherto. Why is this, and what conditions of molecular structure are requisite for the production of a useful fiber? We regret that the limitations of space prohibit a detailed discussion of these and accessory questions, and permit only the baldest statement of our conclusions.

We picture a perfectly oriented fiber as consisting essentially of a single crystal in which the long molecules are in ordered array parallel with the fiber axis (Fig. 10). (In actual fibers a considerable number of molecules fail to identify themselves completely with this perfectly ordered structure. The high strength in the direction of the fiber axis and the pliability are

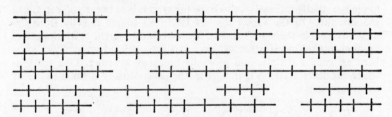

FIG. 10. Lattice from molecules of unequal length

accounted for by the higher cohesive forces of the long molecules and by the absence of any crystal boundaries along the fiber axis. Fiber strength should depend upon molecular length, and recent work by Dr van Natta in this laboratory indicates that it is not possible to spin continuous filaments from the polyester of hydroxydecanoic acid until its molecular weight reaches 7,000. The property of cold drawing does not appear until the molecular weight reaches about 9,000. From these and other facts we conclude that a useful degree of strength and pliability in a fiber requires a molecular weight of at least 12,000 and a molecular length not less than 1,000 Å. (The limits for polyamides may perhaps lie at somewhat lower values.)

Besides being composed of very long molecules, a compound must be capable of crystallizing if it is to form oriented fibers, and orientation is probably necessary for great strength and pliability. Linear condensation polymers are quite generally crystalline unless bulky substituents are present to destroy the linear symmetry of the chains; addition polymers, especially those produced from vinyl compounds, are more rarely crystalline. Possible reasons for this have already been discussed in part VIII. Three dimensional polymers are obviously unsuited for fiber orientation, and synthetic materials of this class are besides invariably amorphous. Glyptal resins belong to this class. It has been proposed to use glyptal resins for the production of artificial silk, but our attempts to carry out the proposed process led to exceedingly fragile (though lustrous) threads which showed no signs of orientation when examined by X-rays.

Although one will not expect oriented fibers to arise by any process of spontaneous crystallization under ordinary conditions, the phenomenon of cold drawing is perhaps only accidentally associated with the capacity to yield oriented fibers; it apparently requires a certain degree of softness and suppleness in the molecules; and its mechanism is doubtless analogous to that involved in the mechanical orientation of cellulose preparations. (Perhaps however, in the unoriented polyesters the molecules are in spiral form and become extended during orientation.)

It has been suggested repeatedly by Staudinger that the great sensitivity of cellulose and rubber to degradation by heat and by certain reagents is due to the fact that the upper limit of thermal stability for the very long molecules lies close to room temperature. It is of interest in this connection that the synthetic linear superpolyesters, in spite of their very high molecular weights, are formed at 200 to 250°C, and that they show no signs of being degraded by repeated exposure to elevated temperature.

Of the remaining papers in the series, only one, number XVIII, is of special interest in the present context. This dealt with the detailed relationship between properties and molecular weight of the polyesters of ω-hydroxydecanoic acid, represented by the general formula,

$$HO-[-(CH_2)_9-CO-O]_n-(CH_2)_9-COOH$$

As these polyesters are typical of those in which the structural unit ignoring the ester linkage is based on a linear aliphatic group—in this case a series of nine successive CH_2 units—their properties are summarized below:

(i) Solubility—For molecular weights less than 10,000, they dissolve rapidly and completely in cold chloroform and benzene and in various hot organic solvents; the highest members show greatly decreased solubility and swell before going into solution.

(ii) Melting point—The melting points increase with increasing molecular weight up to 1,000, the value then being about 70°C, but between 1,000 and 25,000 there is no further increase in melting point.

(iii) Crystallinity—All the polyesters are microcrystalline, the lower ones are waxy and brittle and the higher ones hard, horny and very tough.

(iv) Spinnability—Polyesters above a molecular weight of 10,000 may be spun into fibres; the initial filaments are opaque and fragile and on stretching become pliable and very strong; X-rays show that fibre orientation takes place on stretching. Below a molecular weight of 5,700, no fibres are obtained; at 7,300 filaments are obtained but these can not be extended; at 9,330 stretchable but very weak fibres are produced.

Although the total number of polyesters prepared by Carothers was large, they were predominantly based on aliphatic bifunctional systems, both as regards the alcohol and acid components. As early as the second paper of the series,[1] an extensive series of phthalate esters was described. These were based on phthalic acid, an aromatic dicarboxylic acid of structure

and glycols belonging to the class, $HO-(CH_2)_n-OH$, where n had values of 2, 3, 6 and 10. The products obtained all differed from the aliphatic polyesters in that they were not crystalline. The authors commented: 'The phthalates and fumarate evidently have less symmetry than the saturated straight chain purely aliphatic type.' We shall see later that this was to prove an important clue to Whinfield.

All told, there are fifty-two United States patents in which Carothers appears as the sole or joint patentee. They are all dated between March 1934 and January 1940. Only two, one dealing with polyesters of trimethylene carbonates[2] and the other with polyesters of dibasic acids,[3] notably those between trimethylene glycol and succinic and sebacic acids, refer to polyesters and then not specifically to fibre-forming properties. It appears that the possibility of

[1] W. H. Carothers and G. A. Arvin, *Journal of the American Chemical Society*, *51*, 2560–70 (1929).
[2] U.S. 1,995,291, 26 March 1935.
[3] U.S. 2,012,267, 27 August 1935.

commercial development of fibre-forming polyesters was not seriously contemplated.

Once the polyamides had been examined and found to possess, in addition to enhanced fibre-forming properties compared with the aliphatic polyesters, the necessary high melting point and chemical stability, a strong development effort was launched, especially on the polyamide of adipic acid, $HOOC-(CH_2)_4-COOH$, and hexamethylene diamine, $H_2N-(CH_2)_6-NH_2$, which we now know as Nylon 66. The first patent[1] on these substances was granted on 20 September 1938 and full scale production began within a year.

There remains the interesting question of why Carothers, having set out in unequivocal terms the symmetry, crystallinity and molecular weight criteria for fibre-forming polymers, did not examine the polyesters of symmetrical aromatic dicarboxylic acids with ethylene glycol, or those from the intramolecular condensation of the aromatic parahydroxycarboxylic acids. The commonly made suggestion formally recorded by Izard[2] is that polyesters were simply pushed aside in the hustle and bustle of the Nylon development, but this seems a rather superficial judgment in view of the fact that Carothers, himself, was not apparently heavily involved in the development programme.

It cannot be validly argued that terephthalic acid was not available, or that it was not tried because it was at best a laboratory reagent only, since many of the other compounds used by Carothers were also in this category. Moreover, the existence of a six-membered ring in a linear polymer chain could not have been a serious deterrent, since cellulose contains such a structure. One of the possibly significant factors is the order in which the work was published and which, presumably, corresponds roughly with the order of its execution. The experiments using phthalic acid were among the first carried out well before all the necessary criteria for fibre formation had been established and in a period in which the emphasis was more on fundamental research without a deliberate eye to possibly interesting commercial materials. Another point is that by the time these criteria were settled, polyesters were firmly associated with low melting points and a lack of hydrolytic stability. The information which then existed could not, in itself, have led to the proposition that substitution of an aromatic for an aliphatic group in the chain would necessarily over-

[1] U.S. 2,130,523, 20 September 1938.
[2] Izard, op. cit.

come these deficiencies, especially if, as seems likely, the lack of adequate hydrolytic stability was thought to be primarily associated with the presence of the ester linkage rather than with the nature of the structural unit. Whinfield has himself confirmed that the vast improvement in these properties in polyethylene terephthalate compared with the aliphatic polyesters of Carothers could not be and were not predicted by him.[1]

There are a few other points of possible relevance. Terephthalic acid, though simple in molecular constitution, is a rather intractable substance, being sparingly soluble in most solvents, subliming without melting at about 300°C and showing some capacity to form polymerized anhydrides. The inclusion by Carothers of phthalic acid along with the aliphatic dicarboxylic acids cannot be construed as a systematic attempt to examine the aromatic dicarboxylic acids in general, but probably reflects the fact that, of all the dicarboxylic acids tried, it was probably the one most readily available. Furthermore, there is no evidence in the series of papers or elsewhere that there was any attempt to go back and fill in the gaps that may have been left in the earlier experiments. It is possible that after polyamides were found to have the properties of a desirable synthetic fibre, Carothers simply lost interest, but this could scarcely have happened to all the members of the group engaged on this work.

Finally, we may properly ask why a period of nearly a decade elapsed before Whinfield and Dickson tried their particular experiment, and whether or why the same idea might not have occurred to someone else in the interval. The papers were, after all, published systematically in a leading international journal, they were clearly written and were supplemented by a major review of the field published as paper X of the series in 1931.[2] In addition, Carothers visited the United Kingdom in 1935 and contributed a paper[3] to the Discussion of the Faraday Society in the presence of many of the leading scientists interested in this field. Whinfield was present and recalls meeting Carothers on that occasion. Part of the explanation may lie in the fact that, though Carothers was a shy, reserved man not at his best in the public lecture room, he was, nevertheless, a commanding figure in this area of chemistry; and part may be attributed to the

[1] J. R. Whinfield, *Endeavour*, *11*, 29–32 (1952); also personal communication with the author.

[2] W. H. Carothers, *Chemical Reviews*, *8*, 353–426 (1931).

[3] W. H. Carothers, *Trans. Faraday Society*, *32*, 39–49 (1936).

widespread recognition that the du Pont Company was very experienced in the development of new products and processes, including modified cellulose fibres. If the du Pont Company was not actively following up polyesters with a view to commercial exploitation, others might reasonably have inferred that the prospects were not attractive.

Whinfield and The Calico Printers' Association

Calico printing is concerned quite generally with the production of designs on any fibrous material without recourse to embroidering or appliquéing. Though its origins are in some doubt, printing with blocks was practised in China and India from earliest recorded times and spread through Egypt and the Levant. The secrets of the trade were brought to the United Kingdom by refugee silk printers from France, the first works being established in 1690 on the banks of the Thames near Richmond. Initially, the process was carried out by hand using wooden blocks on which the designs were cut, a separate block being used for each colour in a design. The output was, in consequence, small.

In 1785, to meet the needs of the much larger volume of cotton cloth then being produced as the result of the introduction of steam power and the several cotton spinning and weaving inventions which marked the latter half of the eighteenth century, the wooden blocks began to be replaced by engraved copper rollers. Instead of the cloth remaining stationary, it was now passed under tension between a central drum and the printing rollers which turned it. Each roller was furnished with a separate colour and, provided the fit of the pattern could be maintained, high rates of continuous production of multi-coloured printed designs could be obtained. This invention of roller printing, due to a Scot named Bell, was first introduced at the works of Livesey and Hargreaves near Preston, Lancashire, and rapidly spread to other works in Lancashire and Scotland. By 1840, there were 93 firms in Lancashire with 435 machines and in Scotland 70 firms with 75 machines. Production grew rapidly and from 1893 to 1898 the average yearly production of printed calico for export alone was 960 million yards, a figure which had doubled in the preceding thirty years. In this period of expansion, competition had become destructively keen, prices were being cut and standards of work were declining. A few firms were operating profitably, but many were losing money and in the 1890's consolidation of the small individual firms seemed an obvious answer.

The Calico Printers' Association Limited was formed on 8 November 1899 with an issued capital of £8·2 million as the result of the amalgamation of 46 firms of printers, 32 in England and 14 in Scotland, comprising some 85 per cent of the British calico printing industry and involving a total of 830 printing machines. A number of the constituent firms owned spinning, weaving and dyeing plants which also became part of the Association. It was one of several such amalgamations which took place at the turn of the century, but is the only one remaining in operation today. The Association was not, as J. W. Mitchell claimed in his Jubilee Lecture to the Society of Chemical Industry,[1] in any sense a research association (nor was there ever a body referred to by him as the British Calico Manufacturers' Research Association), but was formed initially for wholly commercial reasons. Among the several advantages which the original directors considered the amalgamation would produce was, however, the following:

(4) That the magnitude of the operations will practically ensure to them the first offer of all new inventions and discoveries relative to the trade and also the best productions of the designer and engraver.[2]

The early years of the new body were difficult ones arising from the large number of long established, highly individualistic firms entering the amalgamation, the unwieldy initial board of directors numbering 84 and the existence of a large body of 128 vendors who retained the management of their branches for five years. In brief, responsibility and control were not in the same hands. After a stormy Annual General Meeting in 1902, which had to be held in the Free Trade Hall, Manchester, in order to accommodate the large number of shareholders who wished to attend, a committee was appointed to recommend an efficient scheme of organization. The report of this committee, which was adopted later the same year and formed the basis of the 1902 Articles, provided for a board of directors of not less than six or more than nine, and for the setting up of sub-committees to deal with such matters as works' production, design, style, prices, trading, cloth buying, drugs, stores, coal and the concentration of

[1] J. W. Mitchell, *Chemistry and Industry*, 29 May 1965, pp. 908–35.
[2] *Fifty Years of Calico Printing*, The Calico Printers' Association Limited, Manchester, 1949.

facilities. The report also pointed to the need for a statistical department to collect and collate information if proper control was to be exercised.

During the period of reorganization, Sir William Mather prepared a separate memorandum on the establishment of a research department 'equipped with necessary appliances for research and experimental work, conducted by the ablest and best trained chemists, specially qualified to pursue investigations in which chemical processes and mechanical and electrical appliances are involved.'[1] Although the sponsor of this proposal was not apparently a member of the Board under the new 1902 Articles, his proposal led in 1906 to the establishment of the C.P.A. Research Department, which Whinfield joined some eighteen years later.

In the early 1920's, the company operated one laboratory as a central research and development establishment. Later, two small laboratories were built at particular works to facilitate attempts to come to grips with problems at or near the production level and to accommodate Whinfield and another officer who had become restive. None was large by present day standards, each having a staff of the order of six persons. Whinfield began in the central laboratory and moved first to one of the small laboratories at which he was not very happy and then, in the late 1920's, to the second at the Broad Oak Works at Accrington as officer-in-charge. During this period, the central laboratory was under the control of L. A. Lantz who had come from Alsace. Its main work at that time was on crease-resistant finishes which had been stimulated by the Tootal patents.

During the period, 1906–49, The Calico Printers' Association took out and maintained 202 patents comprising 68 for mechanical improvements, 116 for chemical processes and 18 for engraving. It is evident from the names appearing on the chemical patents and from their nature that most of them originated from the research department. Excluding that relating to Terylene, Whinfield's name appears on eight patents in the years from 1926 to 1941 and, with one exception, all deal with the treatment of silk, rayon and cotton to achieve some particular property, such as delustering, relustering, imparting crease-resistance, and creating the damask effect. In general, these processes relate to the chemical modification of the fibre or textile surface and many of the other chemical patents are of a broadly similar kind.

[1] Ibid., p. 23.

During the latter half of his period with the Association, Whinfield was on a series of three year contracts and, presumably, this system also applied to other professional staff in the laboratories. It gave limited security over a period and offered no real hindrance, since the chances of anyone getting another job at that time were very poor. Whinfield has recalled that the C.P.A. did not like people to leave and, considering their difficulties, were widely rated as good employers.

The Association suffered acute difficulties between the two World Wars, largely as a result of foreign competition and the depressed economic conditions which persisted during much of this period. The twenty-nine print works in operation in 1918 had by 1939 been reduced to eleven, and during the whole of the period from 1929 to 1947 no ordinary dividend was paid. In spite of these problems, the research and development effort was sustained with the object of achieving better quality at lower cost and to this end of searching for new methods of printing, dyeing, engraving and finishing.

The difficulties which beset The Calico Printers' Association between the Wars could scarcely have created a climate favourable to research and development, though there were clearly those who believed that efforts in this direction would contribute significantly to the firm's survival. One such point of view is vigorously set out in a little book of 'letters' which appeared in 1929 in the following terms:

The C.P.A. needs and must have its Pioneers and Bold Explorers.—In an industry which has peculiar need of them.—It cries out for them—it begs for them beseechingly:—they may bruise and knock some holes into your rigid system—but, do not be afraid of them. They will not fear you. Strong, skilled men inside the C.P.A. will bring you strength. (It is an industry in which the bold individualist can force his recognition) . . . Free, strong, skilled men elsewhere might gain the power to draw your teeth.[1]

It is not clear, however, whether the expressed need was for technical or commercial pioneers or for both.

Under these circumstances, it is pertinent to ask what sorts of research programmes were devised for these laboratories and on what detailed basis they were chosen? Whinfield recalls that part of the task of his unit at Accrington was to deal with local problems which

[1]*Some Thoughts on the C.P.A.—A Plan for Adventure and a more sporting spirit*, Old Calico Printer (pseud.), George Falkner and Sons Ltd., Manchester, 1929, p. 42.

arose on the plant or in the currently operated processes, but in a traditional industry operating under economic duress this was not often a fruitful source of work. The real issue seems to have been to find something worth doing which would also prove acceptable to the management. Faced with this problem, Whinfield made the proper and intelligent choice of placing a proportion of his effort on the raw materials which his employer bought from outside sources; of these, starch was an important one. This choice had the added advantage that starch was related chemically to cellulose.

Some time in 1935, though the date cannot now be set with any certainty, in connection with this work on starch, Whinfield recalls that he had gone into Manchester one afternoon to look up some references in the public library, the facilities at Accrington in this direction being very rudimentary. Wholly by accident in looking through a number of the *Journal of the American Chemical Society*, he came across one of Carothers' papers. Though he is unable now to identify the particular paper, it seems likely that it was the group of papers numbered XI to XVI inclusive, which were published as successive contributions in the issue of volume 54 of 6 April 1932 from pages 1557 to 1599. Of this group, papers XII and XV look the most favourable candidates.

To Whinfield with his extensive background in cellulose, textile and starch chemistry, it is not difficult in retrospect to imagine the impact which these papers had. As the quotation from Paper XV given above[1] shows, the relationships in structural type between cellulose and silk and synthetic polyesters, amides and anhydrides are spelled out in unambiguous detail. It has also to be remembered that at this time wholly synthetic fibres as we know them today did not exist, and apart from the naturally occurring ones, the only textile fibres in common use were derivatives or reconstitutions of cellulose.

This introduction led Whinfield to study all of Carothers' papers, especially the series on polyesters and polyamides. An early general impression that this created was the recognition of a high degree of rationality in Carothers' work and an appreciation of the advances in knowledge which this synthetic approach had generated. A second result was that in the late 1930's Whinfield began to carry out a few *ad hoc* experiments on polyesterification in a simple way, but, as he had neither a molecular still nor any X-ray facilities, these experiments probably did little more than sustain his interest and contribute

[1] See p. 65.

to the general feel for reactions of this kind. These steps led him ultimately to form a specific hypothesis which was largely described in a lecture[1] Whinfield gave in 1943 before the Terylene discovery was public knowledge, and is further supported by subsequent papers[2] and quotations recorded elsewhere.[3]

This hypothesis stemmed from two main sources, one arising from Whinfield's long acquaintance with cellulose and the second from his reading of Carothers' papers. By the 1930's, it was well established, largely as a result of X-ray studies, that natural fibres such as cellulose and silk were macromolecular structures, essentially linear in type and were definitely microcrystalline. Carothers had shown clearly with polyamides, polyesters and some other types of condensation polymers that linear molecular chains, microcrystallinity and high molecular weight (then interpreted roughly as in excess of 10,000) were the essential prerequisites for a synthetic fibre-forming polymer, and that crystallinity in these materials was closely associated with symmetry in the molecular chain.

Carothers had prepared polyesters from simple, straight chain, aliphatic acids and alcohols in which the groups constituting the molecular chain between the ester linkages were, for the most part, symmetrical, successive $-CH_2-$ groups. He had also made polyesters from the unsymmetrical phthalic acid and found them resinous, non-crystalline, non-fibre-forming polymers. There are three isomeric dicarboxylic acids based on benzene as follows:

phthalic acid   isophthalic acid   terephthalic acid

[1] J. R. Whinfield, *Chemistry and Industry*, *21*, 354 (1943).

[2] J. R. Whinfield, *Nature*, *158*, 930 (1946); idem, *Endeavour*, *11*, 29 (1952).

[3] *Landmarks of the Plastics Industry*, Imperial Chemical Industries Limited, London, 1962, p. 91.

The critical step in Whinfield's thinking was the recognition that if terephthalic acid was used in place of phthalic acid, the necessary symmetry condition for microcrystallinity and fibre-forming properties should be met, provided a polymer of sufficiently high molecular weight could be obtained. There were, however, two other properties which could not at that time be predicted with certainty: (i) what the melting point of the polymer would be; and (ii) what its chemical stability would be. Carothers' aliphatic polyesters had failed on both these counts. That the terephthalic polyester might have a higher melting point than an aliphatic polyester of about the same molecular weight was not wholly unreasonable on purely chemical grounds, but, since the chemical stability of a fibre is a function of its chemical constitution and of the way in which the linear molecules are packed together, no effective prediction in respect of this property was at that time possible. It is, however, fairly certain that Whinfield did not seriously speculate about these two properties.[1]

Perhaps the most striking thing about the proposal to react terephthalic acid with ethylene glycol in terms of this formulated hypothesis was that the very first experiment was successful. At the time, early in 1941, Whinfield had two principal assistants, one of whom, J. T. Dickson, carried out the experiment. Whinfield has spoken of the occasion in these words:[2]

It was Dickson who did the first terephthalic acid-ethylene glycol condensation. I think he merely used an oil bath at about 200°C to drive off the excess glycol after the initial reaction and to force the polymerization to completion. Anyway, he presently came running to me to say that the whole mass had suddenly set solid at this temperature.

This was, to me, quite an unexpected piece of good fortune, but I was equally delighted to notice that the solid mass was opaque, a fact which strongly suggested microcrystallinity. We cautiously raised the temperature until at about 260°C remelting occurred. After some hours we ended up with a very discoloured polymer which, however, showed a feeble though quite definite tendency to draw.

From this point, progress in overcoming the discoloration problem and in assessing the hydrolytic stability of the drawn fibre was rapid. Discoloration was attributed to side reactions occurring because of the presence of impurities in the terephthalic acid, so that the first task was to purify the acid reactant. This was approached in two

[1] Ibid., p. 91.
[2] Ibid.

ways: (i) by making the potassium salt of the acid with the object of purifying this by recrystallization and finally recovering the acid in pure form; and (ii) by esterifying the acid to dimethyl terephthalate, purifying this by distillation and, if necessary, then recovering the pure acid. In the event, the first method was not very successful, but the second was and led to the ultimately preferred method of carrying out the polyesterification.

There are, in principle, several ways in which simple esters can be made and Carothers had shown that these were also generally applicable to polyesterification. Apart from the direct reaction between an acid and alcohol

$$R_1-COOH+R_2-OH \rightarrow R_1-CO-O-R_2+H_2O$$

which had been used in the first experiment, there is, among others, the possibility of using ester interchange of the type

$$R_1-CO-O-R_3+R_2-OH \rightarrow R_1-CO-O-R_2+R_3-OH$$

Having made the dimethyl ester of terephthalic acid in order to purify it, Whinfield and Dickson used this method to prepare the polyester according to the scheme

$$CH_3-O-CO- \bigcirc -CO-O-CH_3+ HO-CH_2-CH_2-OH$$
dimethyl terephthalate   ethylene glycol

$$\downarrow$$

$$CH_3\Big[-O-CO-\bigcirc-CO-O-CH_2-CH_2-\Big]_n OH +CH_3OH$$
polyethylene terephthalate   methyl alcohol

This was subsequently recorded as the preferred method in the principal patent[1] and remains the basis for the industrial process today. It not only overcame the discoloration problem, but also improved the fibre drawing capacity.

One of the earliest experiments performed as soon as a fibre could be obtained was to immerse it in both the stretched and unstretched conditions in a molar solution of caustic soda. Immediately, it became evident that there was, in contrast with the aliphatic polyesters, no problem of hydrolytic stability. Moreover, it was shown that the unstretched fibre accepted dyestuff to a much greater extent than the stretched fibre. These very limited preparative and assessment experiments were simple in conception, they involved virtually nothing in the way of special apparatus and together they occupied a very

[1] British Patent, 578,079, 14 June 1946.

short period of time. For all this, they established virtually all the essential properties of a promising textile fibre and an effective means by which it could be made.

The application for the patent[1] which was ultimately granted in 1946 was made soon after the experimental work was done early in 1941. The delay in granting the application and the reasons for it will be discussed in the next section. There are a number of important and rather unusual features about this patent. The first is that it was in the names of J. R. Whinfield and J. T. Dickson, but was not formally assigned to or taken jointly with The Calico Printers' Association Limited, in contrast with most of the other, earlier patents dealing with textile finishing which bear Whinfield's name. Indeed, in neither the specification nor the abstract of the patent[2] does the name of the Association appear at all. The second point is that it was a rather limited patent covering terephthalate esters made from glycols of the type, $HO-(CH_2)_x-OH$, and terephthalic acid and the low aliphatic esters of the acid, such as dimethyl terephthalate. Two methods of preparation were described, namely, direct esterification using hydrogen chloride as a catalyst and ester interchange using alkali metal catalysts. The need to use a glycol to acid ratio in the range, 1–6 to 1, was also stipulated. The preferred method was described in terms of the ester interchange route using a glycol to ester ratio of 10 : 4 in the presence of 0·003 g sodium with the conditions specified in three stages:

(i) heating for three hours at 197°C in an inert atmosphere with continuous removal of the evolved methanol;

(ii) heating for thirty minutes at 280°C at ordinary pressures;

(iii) heating for ten hours at 256°C in vacuo.

The product obtained was described as a white powder.

No concerted attempt was apparently made at the time to undertake peripheral patenting to cover alternatives and variations of the central theme which might reasonably be expected to produce materials of similar properties. Some of these were, however, covered by a patent[3] in 1948 and, in the meantime, du Pont had after 1944 thoroughly exhausted many other possibilities stemming from the initial discovery and the hypothesis on which it was based.[4] In the

[1] British Patent 578,079, 14 June 1946.
[2] Chemical Abstracts, 41, 1495 (1947).
[3] British Patent 604985, 14 July 1948.
[4] Izard, Chemical and Engineering News, 32, 3724–28 (1954).

event, none of the alternatives turned out to be quite as good as polyethylene terephthalate, and in retrospect it seems that Whinfield and Dickson were rather fortunate not to have had their patent circumvented.

The quiescent period

Towards the end of 1941, Whinfield left the employment of The Calico Printers' Association to join the Ministry of Supply. This move had no connection with the Terylene discovery, but was the outcome of his desire to play a more active role in the British war effort. His work at the Ministry was primarily that of advising on the suitability of cotton and other textiles for a variety of service uses and drew more upon general textile experience than particular chemical knowledge. The Terylene patent application had been submitted and there for the moment the matter rested.

During the war years, all patent applications received at the British Patents Office were, as a matter of routine, examined by the appropriate Ministry or other authority in order to determine if there would be any national disadvantage in making the contents public in the normal way. The Whinfield and Dickson patent application in due course thus found its way to the Ministry of Supply to be vetted by the officer charged with this task who happened to occupy an office adjoining that used by Whinfield with whom he was acquainted. Not unnaturally, he consulted his neighbour. The application was declared secret in accordance with the normal practice at that time and, though Whinfield had no formal responsibility for this kind of activity or for the decision that was taken, it appears that some special interest may have been taken by the Ministry in the invention.

The Ministry of Supply arranged for some larger samples of polymer to be made and spun, this work being done by Dr D. V. N. Hardy at the Chemical Research Laboratory (afterwards the National Chemical Laboratory and now disbanded) of the Department of Scientific and Industrial Research at Teddington, but this does not appear to have added much to the state of knowledge at the time. The object was primarily to obtain larger samples for detailed assessment of fibre properties. In 1943, a sample was submitted by the Ministry to Imperial Chemical Industries Limited for preliminary evaluation. When it became evident that the invention had considerable value, a meeting between the Ministry and I.C.I. was arranged to discuss further developments. Whinfield recalls that he was invited to accompany the official Ministry representative on an informal basis so that he could, if necessary, assist his official colleague on any matters of detail. The latter fell ill at the last moment, so that Whinfield became the de facto Ministry representative in respect of his own invention. The I.C.I. representatives were J. C. Swallow, whose role in the polythene story has already been described, and R.G. Heyes. The principal reason why I.C.I. were approached in the first instance lay in the fact that they had had experience in melt spinning, a technique which the development of polyethylene terephthalate was likely to require. As a result of this meeting, negotiations[1] were begun at the end of 1943 between I.C.I. and The Calico Printers' Association Limited.

Superficially, there may appear to have been an element of doubt about the ownership of the patent. The Association clearly had a claim to ownership in common law, if not in the terms of the contract of employment which Whinfield had at the time the invention was made, though it is apparent that the C.P.A. may have treated this patent rather differently from some others which had a more direct bearing on its immediate technology.

Whinfield is, however, quite positive that there was never any doubt that The Calico Printers' Association owned the master patent. The fact that an employer's name does not appear as a joint patentee, or that the patent is not publicly assigned to him, does not abrogate his rights of ownership. Some firms which do not wish to disclose that they are working in a particular field deliberately adopt the practice of not having their names publicly associated with particular patents, though this does not appear to be the reason in this case.

By early 1947, negotiations between I.C.I. and C.P.A. were completed with the former obtaining world manufacturing rights outside the United States. From this point onwards, The Calico Printers' Association played no further role in the United Kingdom except that of receiving royalties which, over the life of the patent, have run into several million pounds. This has enabled the Association to

[1]One version is that C.P.A. approached I.C.I. in the first instance (see, for example, *Chemistry and Industry*, 111, (1955), but the correct sequence of events is that I.C.I. asked the Ministry of Supply if they might approach the owners of the patent; this was agreed to, since the Ministry had, by this time, recognized that the invention could not be developed quickly enough to be of any use in the war effort.

diversify into areas of business in which it had not previously operated.

In the late summer of 1944, a limited amount of information on polyethylene terephthalate polyester fibres was communicated by Imperial Chemical Industries to the du Pont Company in the United States under the terms of a then existing exchange agreement between these two companies. This news must have been a rather bitter pill for du Pont as the founding father of polyamides of the Nylon-type and the sponsors of Carothers' work on polyesters. As a result of receiving this information, du Pont initiated almost immediately their own research programme on polyester fibres and 'by what turned out to be nearly identical procedures, work in the U.S. also resulted in the synthesis of a high melting, orientable, strong polyester, polyethylene terephthalate'.[1] In due course, du Pont negotiated their own agreement with The Calico Printers' Association for manufacturing rights in the United States and the development of this invention on the two sides of the Atlantic thus proceeded more or less independently.

In these negotiations Whinfield had no part, but continued as an officer of the Ministry of Supply. He had made up his mind after the war not to return to The Calico Printers' Association, though he had been invited to do so, and was about to become a permanent civil servant when, in 1947, he joined I.C.I. From that time, he again began to play a significant part in the development of the invention for which he and Dickson, who had also joined I.C.I., were initially responsible. Early in 1947, Whinfield negotiated with The Calico Printers' Association an agreement under which he was to receive a proportion of royalties accruing to the Association with a ceiling of £600 per year.

Development

The development of an invention of this kind may be considered under four main headings as follows: (i) raw material supplies; (ii) process development and design; (iii) product utilization; (iv) market development. These are not wholly separable entities and the total task is to bring them into a single co-ordinated framework. The process itself was a comparatively simple one, the main problems being those arising from the decisions on the scale of opera-

[1] Izard, op. cit.

tions and the common problem with polymer production of achieving a uniform and reproducible product.

There already existed in both I.C.I. and du Pont considerable experience in carrying out polycondensation reactions on an industrial scale arising from the earlier development of Nylon. The principal tasks were posed, therefore, by certain differences in the detailed chemistry of the new process. In both polyamide formation from adipic acid and hexamethylene diamine and in making the polyester from terephthalic acid and ethylene glycol, the stoichiometric requirements are for a molar ratio of 1 to 1 of the two reactants. In the case of the polyamide, this presents no difficulty, since the acid and diamine are capable of salt formation by reaction with each other to form a crystalline product which contains conveniently one mole of each reactant. Heating of this salt brings about condensation by the elimination of water to form the amide linkages; the equilibrium for this reaction favours the formation of the amide linkages, even in the presence of large amounts of water.

In the polyester case, the chemistry was different to the extent that there was no simple way to start with an exact molar balance between the reactants except by careful weighing. Moreover, the equilibrium for the polycondensation reaction was unfavourable for the formation of the polyester, and the by-product, water in the case of direct esterification and methanol or other low aliphatic alcohol in the case of the ester interchange route, had therefore to be removed as fast as it was formed. In addition, ester interchange was faster than direct esterification and, as a result, if the reactants were used initially in mole to mole ratio, undesirable competing reactions occurred. The answers to these difficulties lay in the use of a large excess of glycol at the beginning in order to form first a glycol ester called bishydroxyethyl terephthalate. By subsequently raising the temperature and applying a vacuum to aid in the evaporation of the glycol, ester interchange took place rapidly and displaced the equilibrium in the direction favouring the formation of the polymer. The need to start with excess glycol and to carry the process through in this way was, it will be recalled, the preferred method given in the principal patent, though it is doubtful if the theoretical significance of this procedure was fully appreciated at the time the application was made.

The early emphasis on product utilization was in terms of the fibre-forming properties of polyethylene terephthalate. There were, in general, two forms in which such a material could be presented for further processing; one was continuous filament yarn and the other staple fibre. The technology in both situations was known and the special requirements for Terylene were largely adaptations to suit its particular properties. In filament yarn production, the molten polymer freed of moisture was extruded at a high pressure through a spinneret and the solidified individual filaments wound on cylinders as undrawn yarn. The yarn was then taken to draw-twist machines where it was stretched to several times its original length and wound off on bobbins. The development of methods for making yarn in several different gauges (deniers) and by suitable modification of the processing technique for producing yarn of normal strength and high strength, the latter for industrial uses, did not involve any radically new departures from the then known technology. In addition, bright and delustred yarns could be made and, though dyeing for a time presented some problems, these were soon largely overcome.

Staple yarn production also followed the general lines of the then existing technology. In this case, the extrusion of the molten polymer was carried out on a coarser scale and the filaments brought together to form a rough tow which was then drawn and subsequently mechanically crimped and heat set. The tow was cut into specified lengths depending on the particular textile process for which it was intended. In neither of these cases could it be said that the conversion of raw polymer to usable basic material presented any insuperable difficulties, though the work in these stages necessarily took some time.

Market development was initially directed to two broad areas, textiles and industrial products. In these, use was being made of all or some of the seven outstanding properties which the fibre and materials made from it possessed. These were: (i) warmth to the touch; (ii) low stretch; (iii) crease resistance and a capacity to retain shape whether wet or dry; (iv) ease of washing and rapidity of drying; (v) high tenacity and high resistance to abrasion; (vi) moth and mildew proofness and immunity to bacterial or insect attack; and (vii) good resistance to sun and weather. Market development largely consisted of finding those applications for which Terylene was uniquely suited and assessing the extent to which textiles made wholly from the synthetic or in blends with natural fibres might successfully invade the appropriate markets. Sampling of the market was an obvious approach, and for this significant quantities of material were required.

Of all the factors influencing the development and commercial exploitation of this invention, none was of more importance than the supply of chemical raw materials. Basically, these were three in number, terephthalic acid, ethylene glycol and methanol. The last two were well established industrial chemicals, though considerable expansion in the scale of production of ethylene glycol was needed. This substance is most conveniently made from ethylene oxide, the availability of which depends ultimately on supplies of ethylene. Before Terylene exerted demands for terephthalic acid, this was not an organic chemical of commerce. Indeed, at an early stage the Ministry of Supply was advised on good authority that the prospects of making terephthalic acid on an appropriate scale at an acceptable cost were very remote. This thinking persisted strongly in the immediate postwar years and could easily have led to the total abandonment of the project in 1949. The prospective market for the polyester envisaged at that time was too small for the petroleum firms to be willing to erect a plant specifically to make p-xylene required and too large for I.C.I. to make it themselves from sources other than petroleum. One of the factors which may have influenced I.C.I. against abandonment at that time was the knowledge that du Pont were going ahead.

The solution of this raw material supply problem was to be found in the concurrent developments within Imperial Chemical Industries of a major petrochemical complex at Wilton, Yorkshire. Until this was well advanced, Terylene could only be produced in comparatively small quantities, sufficient however, to provide an adequate test of the market. At the end of 1949, a pioneer polymer plant was established at Huddersfield, Yorkshire, and a pilot spinning plant at Hillhouse near Fleetwood, Lancashire. The choice of these particular sites was primarily due to the existence of other manufacturing facilities which the company had at these places. By 1952, production from these pilot plants had reached 600,000 pounds per annum and by 1955 this figure had increased to nearly two million pounds.

Almost from the outset, it became obvious that there existed a substantial market for this new polyester fibre and in October 1952, I.C.I. announced the decision to construct a large plant at Wilton to make 11 million pounds of Terylene a year. The aim was to have this plant in production at about the turn of the year 1954–5. In the event, production began early in 1955, but the decision to double the capacity had already been taken a year before with this additional output to become effective early in 1956. Construction and production on this sort of scale, once the process had been established at Huddersfield, rested on the construction at Wilton of plant facilities for producing the necessary synthesis streams.

The site of about two thousand acres had been acquired at Wilton in 1946 and in the decade which followed more than a dozen different plants were erected there by six I.C.I. Divisions, the first coming into operation in April 1949. The essential reason for this concentration was that this site was planned as a petrochemical complex utilizing chemical synthesis streams produced from the cracking of petroleum fractions with subsequent processing and purification of the products. Ethylene, a large proportion of which was intended for the manufacture of polythene, also provided the necessary raw material for ethylene oxide and thus ethylene glycol. The other essential starting material, p-xylene, which could be oxidized to terephthalic acid, was also conveniently made within the framework of the Wilton concept. Without such a cognate development, it is doubtful if Terylene production on a large scale could have been seriously contemplated. The p-xylene made at Wilton was, in fact, the first aromatic hydrocarbon to be made from oil in the United Kingdom.[1]

Some idea of the total scale and rate of building and commissioning of chemical plants at Wilton is afforded by the fact that by January 1955 an expenditure of £60 million had been sanctioned, £40 million had already been spent, expenditure was running at £750,000 per month and it was expected that by 1965 the total capital investment on the site would be in the vicinity of £100 million. The first stage of Terylene production to a level of 11 million pounds per year accounted for an investment of about £10 million, a figure which was to be doubled by the end of 1956.

By January 1955, I.C.I. through its subsidiary company, Canadian Industries (1952) Limited, was building another plant at Millhaven near Kingston, Ontario, also with a capacity of 11 million pounds per year, and negotiations had been concluded for the manufacture of Terylene under licence in Italy, France, Western Germany and the Netherlands.[2]

---

[1] *Landmarks of the Plastics Industry*, Imperial Chemical Industries Limited, London, 1962, p. 80.

[2] Most of the detailed figures cited in the preceding paragraphs are based on published information. See, for example, *Chemistry and Industry*, 18 October 1952, p. 1030; 6 February 1954, p. 154; 29 January 1955, p. 111.

Administratively, the development of Terylene within Imperial Chemical Industries was also the occasion for a temporary departure from the customary organizational pattern. In the early stages, the total task was placed under the control of a body called the Terylene Council which co-ordinated and directed the programme utilizing in some cases the facilities and resources of other divisions, among which the Plastics Division was called upon to play an important role. With the establishment of the full scale manufacturing facilities at Wilton, the essential task of the Terylene Council was complete and the whole of the activities were absorbed into the newly created Fibres Division with headquarters at Harrogate, Yorkshire. This division also took over responsibility for Ardil, the protein fibre then being made at Dumfries in Scotland, but which was later discontinued. Noticeably, responsibility for Nylon remained with Dyestuffs Division. In these several moves, Dr A. Caress as Chairman of the Terylene Council and later at Harrogate played a major part. J. R. Whinfield completed his formal associations with this polyester as a director of Fibres Division.

In the United States, du Pont conducted their own development largely independently of I.C.I. and rather more rapidly. This probably reflected the greater experience in melt spinning and the earlier availability of supplies of p-xylene. By 1951, production of Dacron fibre had reached 3 million pounds and by the end of 1953 this figure had risen to 35 million pounds. Plant construction began in April 1951 at Kinston, North Carolina, with the first of six units which began production in March 1953; five more were completed by the end of 1953 involving a total capital expenditure to this point of about $40 million.

On both sides of the Atlantic, it was recognized at a quite early stage that polyethylene terephthalate should also be a useful polymer from which to make film. Because of heavy commitments on fibre development, Plastics Division of I.C.I. did not effectively begin this task until 1951. Again, development of Mylar film by du Pont began rather earlier. The processes that were developed were closely similar and involved three steps as follows:

(i) the extrusion of the molten polymer through a long narrow slot followed by rapid quenching with water to yield an unoriented, amorphous, unstable film of high clarity and gloss.

(ii) the plane orienting of the film which involved stretching it in

7

two directions to improve its strength without significant loss of gloss or clarity.

(iii) heat setting to confer on the film the necessary thermal stability.

This three stage process followed fairly directly from the nature of the polymer and its behaviour with respect to heat. For example, the film emerging from (i) would, without further treatment, begin to crystallize at around 80°C (the glass transition temperature), lose some of its strength and become opaque. The basic facts were well known by this time and the problems were primarily engineering ones. Initially, the demand for the film was depressed because little polymer was available for film making and utilization was restricted to special applications where the low water absorption, good electrical properties and high heat resistance justified its use over the cheaper cellophane and polythene films. du Pont also developed the polyester as a photographic film base under the trade name of Cronar.

An assessment

Here we shall draw together some of the salient features of this narrative and point to some general conclusions. In this, it is convenient to employ the same general time sequence.

The dominant characteristic of the Terylene invention is probably its inherent rationality; there have been few similar discoveries which have been thought out beforehand with such precision. This is not, however, to suggest that what Whinfield did was obvious; far from it, for if this were so, it would be impossible to account for the decade which elapsed between the publication of Carothers' papers and the discovery of polyethylene terephthalate. Moreover, Carothers' work received a good deal of publicity in the 1930's at conferences in U.S. and United Kingdom and in the public press, apart from the publication of the actual papers. A second feature was the characteristics and fortunes of Whinfield, the man, his background and particular experience against which the idea finally crystallized. Here was an example of a university educated chemist working in a highly traditional industry at a time when such persons were comparatively rare and able to bring together profitably elements of science, technology and industrial practice. It seems to have been a long term policy of The Calico Printers' Association to recruit such men into the industry through its research department. The official record[1] of the Association's jubilee in 1949 set this out in the following words:

In addition to its work of regulating and improving works processes and the discovery of new ones, it (the research department) has proved an excellent training ground for University graduates in science and technology, later to be management staff in works and mills.

A third aspect is that despite this policy and despite the fact that the research department dated from 1906, it could hardly be maintained that the Terylene discovery was made under attractive or stimulating research conditions. Possibly because of the economically difficult period for the industry between the two World Wars, the general facilities for research were meagre and the library provisions apparently very poor indeed, at least with respect to the principal international journals in chemistry. And yet, the productivity of the C.P.A. research department judged in terms of the total patenting in these years was creditable. The chemical patents tended to be closely related to improvements in what broadly could be described as finishing processes, but at least the concept of formal invention and the practice of seeking patent protection were well established. Whether these patents were of large intrinsic value is another question, but at least they represent steps of magnitude greater than those of evolutionary change. In brief, the climate for innovation seems to have been one lying somewhere between the evolutionary pattern of a typically traditional craft industry and one of revolutionary innovation involving strikingly new products and processes. In this respect, it was a situation in which the translation of science into the particular industrial practice was not wholly unfavourable.

With respect to the Terylene invention, the science was provided in a circumscribed way by the work of Carothers and the outlook towards usage and application largely by Whinfield in drawing upon his association and experience with Cross and Bevan and The Calico Printers' Association. There may be some validity in the argument that it was not Whinfield's job to invent synthetic fibres, or to carry out research with this as an implied aim, since this was not the central interest of his employer. On the other hand, calico printing at that time involved the use of both natural and derived polymers, and much of the research effort was directed to modifying their properties

[1] *Fifty Years of Calico Printing*, the Calico Printers' Association Limited, Manchester, 1949.

to achieve particular effects. The invention of quite new fibres was at least peripheral to this long established, central activity.

This incidence of peripheral industrial invention has some aspects in common with what has sometimes been described as innovation by invasion for which the usually quoted example is the invasion of the textile industry by the chemical industry by way of the introduction of synthetic materials (including fibres) to the former's operations. The case of Nylon is an example of such invasion, but for Terylene the situation is a little more subtle—a Trojan horse within the gates—but one motivated by external scientific discoveries.

There is also the interesting question of whether Whinfield's external interest in condensation polymers was stimulated by the C.P.A. climate, or whether it was the result of dissatisfaction, frustration or uncertainty. Taking a broad view of the evidence, it is difficult to escape the conclusion that Whinfield felt some degree of uncertainty about his future and the continuity of his employment at a time when his family responsibilities were growing and when other posts in a depressed industry or elsewhere might not have been easy to find. In addition, it seems that he had some dissatisfaction with the limitations imposed by the nature of the industry in which he worked and was, perhaps, unable to obtain that sense of achievement from his work which often proves to be a major satisfaction to many people. While these assessments and judgments, if they are correct, cannot be taken as advocating the deliberate creation of an environment where limited or inadequate facilities and dissatisfactions and uncertainties of a major or minor kind are employed as a means of stimulating inventions, it is difficult to discount entirely the fact that a considerable body of creative and inventive acts has been the product of such circumstances.

Finally, there was unquestionably an element of good fortune in this discovery, not so much in relation to the particular choice of terephthalic acid as such, but rather in that the qualities of polyethylene terephthalate were not superseded by one or other of the several alternative systems which also flowed from Whinfield's hypothesis. Izard[1] has given a comprehensive account of the du Pont efforts to explore these alternatives as soon as the first news of polyethylene terephthalate reached the United States, and there were certainly no preconceived ideas on du Pont's part at that time that Whinfield had necessarily chosen the best system. In retrospect, there

[1] Izard, *Chemical and Engineering News*, *32*, 3724–28 (1954).

does not seem to have been any special case for choosing the terephthalic acid-ethylene glycol system in preference, for example, to hydroxy acids of the type, $HOOC - \bigcirc -(CH_2)_x - OH.$

Although a period of about six years (1935–41) elapsed between the time of Whinfield's acquaintance with Carothers' papers and the carrying out of the critical reaction, this does not necessarily imply that the idea on which it was based required a long period of gestation. Whinfield says that some notes he made in 1935 show that the ideas used later were fully developed at this early stage, and that part of the reason for the elapse of time was due to his interest in the work on starch which fully occupied him. He was also somewhat deterred by the practical requirements, as the Carothers' papers initially led him to believe that a molecular still and possibly X-ray facilities would be required.

The question of whether the same idea occurred to anyone else before the publication of the patent cannot be answered with any certainty. However, in 1945 and without knowledge of the Terylene discovery, Dr Otto Bayer[1] told Whinfield that the condensation of terephthalic acid and ethylene glycol had been attempted in Germany in a sustained effort to circumvent the Nylon patents, but without fruitful result. If, as seems likely, some others had the same idea, part of the success at Accrington was due to the choice of the correct experimental conditions. In common with many other cases, here is the element of chance which is said to favour those with patience and persistence.

In Whinfield's case, persistence did not, however, extend to the reading of Carothers' patents, as distinct from the papers. Whinfield has provided the following anecdote which reveals something of his approach:

Once when I was in the United States, someone said to me, 'I suppose you studied all of Carothers' patents very carefully'.

When I told him I couldn't recall having read any of them, he clearly thought I was a 'nut case' and went around introducing me to everyone as the man 'who never read patents'. He wasn't too wide of the mark, I hate them.

In the period from 1941 to 1947, the incidence of the war was a seriously complicating feature, without which the development of

[1]Dr Bayer, a distinguished German polymer chemist, had discovered diisocyanate polymers in 1939.

Terylene may well have proceeded at a different pace and in an alternative way. It is, perhaps, unfair to question whether the role of the Ministry of Supply was a help or a hindrance and it is difficult to assess accurately the extent to which a greater interest may have been taken in this particular invention because of Whinfield's presence on the Ministry's staff. On most bases of judgment, though this is necessarily with the advantages of hindsight, it is scarcely possible that an invention of this kind made in 1941 could have been developed in time to contribute significantly to the war effort. Had this been attempted, success may have come only at the expense of other programmes of higher priority. Probably the most important contribution of the Ministry in those years, particularly in the early period, was to keep something moving, firstly through the Department of Scientific and Industrial Research and in catalysing the association between the owners of the patent, The Calico Printers' Association, and the potential developers, Imperial Chemical Industries.

The earlier experience with other synthetic fibres, notably Nylon, and the considerable build-up in plastics production generally in the late 1930's and early 1940's largely ensured that Terylene was not called upon to break wholly new ground commercially. Synthetics had by this time begun to be accepted as having virtues in their own right, rather than being looked upon as substitutes for 'real' fibres. In the creation of this climate, Rayon had also played a pioneering role. Somewhat amusingly, Whinfield recalls that in his wartime years in the Ministry, he was not infrequently confronted with the task of examining proposals that various items of equipment should be made of Nylon on the premise that cotton was in short supply. Certainly, the rise of the Nylon stocking to the status of a currency in Europe immediately following the war speaks for itself.

There does not appear to have been serious doubt that the development of Terylene would undermine established markets for Nylon, although in a number of applications a fair degree of substitutability was possible. This reflected the high demand for Nylon and some other synthetic fibres which had persisted for many years, and may also have been related to the fact that, in addition to promoting Terylene and Dacron as such, there was a growing tendency in the textile industry to think in terms of blends of natural and synthetic fibres. While the fine wool producers opposed this approach, at least to the extent of declining over a long period to advertise or promote the use of wool in these mixed fabrics, this does not seem to have had any serious inhibiting effects on the fibre and textile manufacturers.

As far as can be determined, there was not at any time a serious proposition that The Calico Printers' Association would undertake the development of polyethylene terephthalate fibres themselves. This is not surprising since they had no experience in large scale chemical manufacture and, in any event, were not in a satisfactory financial position to diversify in this way. In fact, as events have shown, their merchandising rather than manufacturing facilities have increased in the post war years. Imperial Chemical Industries were not, however, the only United Kingdom concern with which negotiations could have been conducted. Courtaulds, for example, may well have had considerable technical claims, although with the greater breadth in chemical manufactures as distinct from fibre production, the choice of I.C.I. was the more obvious one.

It is often claimed that British firms are slow or less competent in development than their counterparts in the United States and some other parts of the world. In the present study, some comparison is possible between I.C.I. and du Pont starting at about the same time, though du Pont with their substantial research and development effort on polyesters generally stimulated after 1944 probably had, by 1947, some initial advantage. The efforts were largely parallel and the fact that I.C.I. were in full scale production (11 million pounds per annum) early in 1955 and du Pont had reached the same stage in the course of 1953 does not bear out seriously this criticism in this particular example. Moreover, as has been pointed out, large scale production of this product rested heavily on the availability of terephthalic acid in the required quantities and hence on the development of petrochemicals generally. Considering the conditions which existed in Britain immediately after the Second World War, the Wilton conception was by any standard of judgment a bold one. Effectively, the date of disclosure of the Terylene invention can be set as 1944, so that the total period from invention to large scale production was, on the average, a decade. In comparison, the actual laboratory discovery and first preparation of the polymer used in manufacturing the first Nylon took place on 28 February 1935 and the first commercial plant at Seaford, Delaware, with a capacity of 3 million pounds per annum began production late in 1939.[1] International comparisons of this kind need, in any case, to be treated with caution, particularly in

[1]W. F. Mueller in *The Rate and Direction of Inventive Activity*, op. cit., pp. 323–60.

the chemical process industries, where one development often depends vitally on the progress of others.

There is some interest in seeking to compare the costs of research and development and construction of plants for a number of synthetic fibres. The figures cited by Mueller[1] for Nylon, Orlon and Dacron by du Pont are summarized below.

		$
Nylon: (1939)	Research and development to stage of building a pilot plant	787,000
	Pilot Plant	391,000
	Market Development	782,000
	First commercial plant (3 million lb p.a.)	8,600,000
	Total	10,560,000
Orlon: (1950)	Research, development, pilot plant and Market Development	5,000,000
	First commercial plant (30 million lb p.a.)	20,000,000
	Total	25,000,000
Dacron: (1953)	Research, development, pilot plant and Market Development	25,000,000
	First commercial plant (35 million lb p.a.)	40,000,000
	Total	65,000,000

These figures have been gleaned by Mueller from various sources and are not, in any sense, official figures. There are some possible inconsistencies, especially in the breakdown between research and development, pilot planting and market development. For example, for Dacron a figure of $6 to $7 million has also been given as the amount spent before the building of the first commercial plant could begin, and a figure of $65 million for the total expenditure of which $40 million was the plant cost. This appears to leave some $18 million to $19 million presumably for pilot planting and market development. There are clearly some problems in definition and allocation of expenditure, but the figures illustrate the general magnitudes which are involved. The I.C.I. plant figure of £10 million for a production level of 11 million lb per year fits fairly well the du Pont expenditure on plant of $40 million to achieve 35 million lb per year.

Finally, an uncommon feature of general importance arises in this instance because the Terylene discovery was essentially the work of one man. In a recent, unsigned article[2] in *The Times Review of*

[1] Ibid.
[2] *The Times Review of Industry and Technology*, 3 November 1965, pp. 12–15.

Industry and Technology, the Whinfield case is cited in support of the thesis that more adequate rewards for inventors are desirable in Britain, especially where inventions of outstanding commercial importance are made in the course of employment. Part of the problem here arises because of the apparently conflicting judgments under Common Law and the provisions of the Patent Act of 1949.

The general practice is that for employees without specific service contracts the common law provision applies, and inventions made in the course of employment are the employer's property. Where a service contract exists, the same provision is almost universally specifically included. The question which arises in both cases is whether, in the event of such an invention being highly successful commercially, the employer should be required to make a reasonable distribution to his employee-inventor and, if so, what is to be interpreted as reasonable, and how could or should this be embodied in legislation. Limited enquiries reveal that reputable employers attempt without legislative coercion to behave in this spirit, but there are necessarily difficulties in deciding what is a reasonable reward, especially where these negotiations are normally carried out before the full commercial potential of an invention is realized. But as the law and practice currently stand, there can be no automatic reward to an employee-inventor and he must be willing to exert himself in negotiations as a matter of business, if he is to profit financially.

The real difficulty in improving by legislation the negotiating strength of the employee-inventor lies in the fact that to provide in legal terms for general application over the whole realm of science and technology, and to meet the infinite variety in human situations which occurs, would probably pose as many problems as it seeks to solve.

IV

OXYGEN STEELMAKING

Introduction

The generic term, steel, covers a multitude of materials in which the principal component is the element, iron, together with small quantities of carbon and minor amounts of a variety of other elements, such as manganese, silicon, sulphur, phosphorus, oxygen, nitrogen, tungsten, nickel, cobalt, chromium, vanadium and so on. The typical compositions of common steels include, in addition to iron, $0 \cdot 1$–$0 \cdot 74$ per cent carbon, $0 \cdot 01$–$0 \cdot 21$ per cent silicon, $0 \cdot 35$–$0 \cdot 84$ per cent manganese, $0 \cdot 028$–$0 \cdot 055$ per cent sulphur and $0 \cdot 005$–$0 \cdot 054$ per cent phosphorus, but not all these elements have been deliberately added. Mild steel, the principal stock-in-trade of the engineering and constructional industries, contains $0 \cdot 15$–$0 \cdot 25$ per cent carbon.

For statistical purposes, it is often more convenient to define special types of steel leaving the remainder in a general category. In Britain,[1] high carbon steels are defined as those containing not less than $0 \cdot 6$ per cent carbon with less than $0 \cdot 04$ per cent each of phosphorus and sulphur, or less than $0 \cdot 07$ per cent of these two elements taken together. Alloy steels have, since 1960, been defined as those containing at least $0 \cdot 1$ per cent molybdenum or vanadium, or $0 \cdot 3$ per cent tungsten or cobalt, or $0 \cdot 5$ per cent chromium or nickel, or 2 per cent manganese. Most of these high carbon and alloy steels are made either from scrap or from a suitable crude steel stock. In this study, we shall be specifically concerned with the use of oxygen in making the common grades of steel.

The ferrous component in steel is derived from two sources, pig iron obtained from the reduction of ores of iron oxides in a blast furnace and from steel scrap arising from a variety of sources within and without the steel industry. While scrap is almost invariably available in the cold solid state, pig iron may be supplied in the solid state or as hot liquid metal, the latter generally in situations in which pig iron producing facilities are physically linked with those for steelmaking. The relative availability and cost of these ferrous components, which may be alternative or complementary, have a major influence on the choice among the several steel refining processes.

[1] *Iron and Steel Annual Statistics for the United Kingdom*, Iron and Steel Board and British Iron and Steel Federation, 1965, p. vii.

While steel scrap, whether derived from manufactured steel goods or from circulating works scrap, will in general have a composition reasonably close to that required in new production, pig iron may vary considerably in composition depending on the iron ore, coke and limestone employed in the blast furnace feed and the operating practice that is adopted. The compositions[1] of four reasonably typical pig irons are shown in Table 1.

Table 1: Compositions of four pig irons

Element	No. 1 %	No. 2 %	No. 3 %	No. 4 %
Carbon	3·5–4·0	3·0–3·6	3·5–4·0	3·0–3·6
Silicon	2·0–2·5	0·6–1·0	1·5–2·0	<1·0
Sulphur	<0·040	0·08–0·10	<0·040	<0·06
Phosphorus	<0·040	1·8–2·5	<0·050	1·0–1·5
Manganese	0·75–1·0	1·0–2·5	2·5–3·0	1·0–3·0

These compositions fall into two groups in terms of phosphorus content, Nos. 1 and 3 belong to the class of low phosphorus pig irons, whereas 2 and 4 contain substantial quantities of this element. This binary classification primarily reflects the composition of the two types of iron ore which commonly occur in nature.

The aim of any steelmaking process is to bring about a change in composition of a pig iron or mixed pig iron-scrap charge to meet a required specification. It is, however, not always possible to achieve exactly and simultaneously the required change for each of the foreign elements, and it frequently happens that, in reducing the sulphur and phosphorus to the required levels, the contents of carbon, silicon and manganese may fall below the respectively required concentrations. These deficiencies are usually remedied by the addition of carbon in the form of anthracite and the other elements in the forms of ferro-silicon and ferromanganese alloys.

While the rudimentary economic aim in steel refining is to produce steel to a required specification at minimum cost, this sector of the iron and steel industry seldom exists in isolation. The iron and steel industry has long been characterized by a tendency towards integration—backwards through scrap procurement, iron making, the manufacture of metallurgical coke and consequential by-products, the ownership and exploitation of deposits of iron ore, coal and limestone and their transportation, and forward towards the casting,

[1] G. R. Bashforth, *The Manufacture of Iron and Steel*, 3rd Edn., Chapman and Hall, London, 1964.

rolling, shaping and forging of angles, shapes, sections, rods, plate, strip etc., the production of tin plate, the formulation of alloy steels and into various large tonnage, steel-using fields, such as ship-building and armaments. In these circumstances, the factors which may determine the rate and course of technological innovation in steelmaking are not necessarily restricted to this stage in isolation, but may strongly reflect the relationship of steelmaking with both its preceding and succeeding activities.

Few would dispute the central role which steel has played in indus-trialized nations during the past century. Some measure of its current general importance on the world scene is afforded by Table 2 which reviews the levels of steel production in the more important national or economic areas. These data reveal a number of features, notably the steady growth of production in U.S.S.R. in contrast with the quite violent fluctuations in U.S.A., the strikingly rapid regrowth in Japan following a somewhat delayed postwar resurgence of that country's steel industry, the vigorous growth in those countries not separately designated and the general impact of the 1958 recession. Of special interest are the figures for the United Kingdom which, for the decade 1954–63, display the lowest rate of growth and well below that for the world as a whole. It is necessary to treat growth rates over such a period with caution, since it cannot be said that in 1954 all the pro-ducing areas had recovered to an equal extent from the effects of the Second World War and, moreover, this base year is within a period when supply was still catching up with demand in many parts of the world.

The tapering off in the growth during the last three years in Table 2 points to a situation developing in this period when available produc-tive capacity appreciably exceeded demand. Over the ten year period, surplus capacity on the world scene increased nearly fourfold. Until 1957, the surplus was small, some 15–20 million tons, but it rose to 65–75 million tons during the recession of 1958–9, falling to 55 million tons in the boom of 1960 and again rising rapidly to 85 million tons in the recession of 1962–3. An attempt[1] has been made to predict the likely situation in 1970, though any such effort is necessarily fraught with uncertainty, especially with respect to the Soviet bloc. The figures set out in Table 3 point to the expectation of a continuation of some surplus capacity.

[1] *Steel Review*, *37*, 33, January 1965.

Table 2: *World crude steel production*[1]

(Million tons of 2,240-lb. Index 1954=100)

Country or Area	1954		1955		1956		1957		1958 (a)		1959		1960		1961		1962		1963 (c)	
	Ton-nage	Index	Ton-nage	Index	Ton-nage	Index	Ton-nage	Index	Ton-nage	Index	Ton-nage	Index	Ton-nage	Index	Ton-nage	Index	Ton-nage	Index	Ton-nage	Index
U.S.A.	78·85	100	104·50	133	102·87	130	100·64	128	76·12	97	83·43	106	88·64	112	87·51	111	87·79	111	97·56	124
U.S.S.R.	40·78	100	44·56	109	47·93	118	50·24	123	54·05	133	59·00	145	64·26	158	69·63	171	75·10	184	78·93	194
E.C.S.C. (b)	43·27	100	51·94	120	56·06	129	59·05	136	57·26	132	62·35	144	71·91	166	72·34	167	71·85	166	72·05	167
Japan	7·63	100	9·26	121	10·93	143	12·37	162	11·93	156	16·37	215	21·79	286	27·82	365	27·11	355	31·00	406
U.K.	18·52	100	19·79	107	20·66	112	21·70	117	19·28	104	20·19	109	24·30	131	22·09	119	20·49	111	22·52	122
Other	31·42	100	35·81	114	40·96	130	44·23	141	48·21	153	59·39	189	68·33	217	68·61	218	69·54	221	74·82	238
Total	220·5	100	265·9	121	279·4	127	288·2	130	266·8	121	300·7	136	339·2	154	348·0	158	351·9	160	376·9	171

(a) Figures for 1958, which was a 53 week production year, have been reduced to 52 weeks for comparison.

(b) European Coal and Steel Community comprising Western Germany, France, Belgium, Italy, Luxembourg, Netherlands.

(c) Provisional.

[1] *Development in the Iron and Steel Industry, Iron and Steel Board, Special Report, 1964*, H.M.S.O., London, p. 154.

Table 3: World steel capacities and outputs, 1963–70
(million ingot tons)

Country or Group	1963 Capacity	1963 Production	1963 Surplus	1970 Capacity	1970 Production	1970 Surplus
U.S.A.	145	98	47	150	115	35
Soviet Bloc	105	103	2	150–160	145–155	6
E.C.S.C.	85	72	13	110	95	15
Japan	40	31	9	60	50	10
U.K.	29	22	7	35	29	6
Intermediate Producers*	41	37	4	70	65	5
Others	5	3	2	10	7	3
Total	450	366	84	590	510	80

*Includes Canada, India, Australia, Sweden, Austria, Brazil, Spain, S. Africa, Mexico, Yugoslavia.

Superficially, these data suggest that the general situation in respect to the balancing of capacity and production is expected to improve somewhat by 1970, but what is probably more important in assessing the future stability of the market for steel is the capacity available for export in relation to the total volume of the world export trade. In 1963, the total capacity available for export was 120 million tons, compared with a world trade volume of 36 million tons; in 1970, the corresponding figures are expected to be 125 and 45 million tons, respectively. These figures suggest that, with continued growth in demand, the problem of more closely matching capacity and demand may be adequately solved in the long term, but, in the meantime, a potentially unstable situation in the world market for steel will persist. It is against this total background that any aspect of technical innovation in steelmaking needs to be considered.

The international distribution in steelmaking and the comparative freedom with which technical information is widely exchanged within and between countries suggest that it might be worthwhile to make a full, comparative, international study of innovation in steelmaking. This poses considerable problems, not the least of which is the physical task of obtaining and assembling all the relevant information over such a wide compass. Alternatively, we may use the device of choosing one country to study in detail and to relate its performance to the general pattern over the wider area. Great Britain offers a number of special advantages for detailed study. In terms of productive capacity, it is the smallest of the large producing countries; it has a long history of steelmaking based on adequate home resources of coal, but limited supplies of local ore of rather poor quality; in the siting of plants, both inland and seaboard works have been established; it is within and yet outside Europe in its economic relationships; and there has emerged over the past thirty years a number of interesting experiments in the relationships between the industry and the state. An array of factors are thus brought together in one place in Britain in a way which is probably unique.

The title of this chapter 'Oxygen Steelmaking', might imply that steel could be made without the use of oxygen, free or in the combined state. Given a supply of iron ore which could be beneficiated by physical methods to provide a feed stock sufficiently low in impurities such as phosphorus, sulphur and silicon, it would be possible in principle to transform this material into steel by a simple process of direct chemical reduction. In practice, however, this chemically ideal situation is seldom likely to be wholly realized, so that the dual processes of chemical reduction to make pig iron followed by the oxidation of the undesirable impurities to yield acceptable steel have had to be employed. Oxidation in itself does not necessarily require the use of oxygen, either in the gaseous state or combined with other elements—indeed, it is possible to conceive a process for refining pig iron into steel which employs chlorine or other suitable element. But oxygen is one of the most abundant elements in the earth's crust and constitutes about a fifth by volume of the atmosphere. In the latter case, it is available free, the main problems being its low density and the fact that it is mixed with a large amount of nitrogen.

The operation of iron and steelmaking processes require, in general, the use of elevated temperatures, and the creation and conservation of heat are often considerations as important in economic terms as the chemical transformations being carried out. Traditionally, the required heat was obtained by the combustion in air of carbon or carbon bearing materials, but most other elements when they combine with oxygen also do so exothermically. There is, therefore, nothing peculiar about the element carbon as a fuel and many other elements will serve as well, and in some cases better. In this study, we shall not be much concerned with the conventional use of oxygen in air in steelmaking practice, but with intensive use of oxygen itself. Such usage requires that oxygen be available in a fairly pure state at an acceptable cost. It will, therefore, be necessary to trace in some detail the developments that have led to the production of what has come to be called 'tonnage' oxygen.

A background to conventional steelmaking processes

By the beginning of the nineteenth century, the iron trade was well established in Britain. Its products comprised a small quantity of crucible tool steel and blister steel mainly used by the Sheffield cutlers, a fair tonnage of wrought iron in the form of bars and sheets and a much larger amount of pig iron, most of which was utilized in castings. By mid-century the steam railway made it possible to assemble on one site substantially increased quantities of raw materials for iron making, thereby facilitating the construction of larger works and the exploitation of more extensive mineral deposits. Production of pig iron in Britain had risen from 400,000 tons in 1815 to about 3 million tons by 1850, but by then much of it was puddled to wrought iron, hammered and rolled into rails, sheets and bars to meet the growing demands of the railways and the constructional industry.

In 1855, Henry Bessemer took out a patent for the manufacture of malleable iron and steel by blowing air through molten pig iron, though the initial object of his experiments was not to make steel, but a modified cast iron suitable for use in guns and capable of being rifled for rotating shot. His first approach had been to improve the reverberatory furnace used at that time for puddling pig iron into wrought iron by enlarging the grate to obtain a larger fire and introducing additional air to intensify the temperature of the flame impinging on the metal in the hearth. He recognized that impurities were being removed from the pig iron by preferential oxidation and that these oxidation reactions were exothermic; from these conclusions, he formulated the hypothesis that molten pig iron might be converted into malleable iron simply by blowing air through it. His original converter was a refractory-lined cylinder about four feet in height with air nozzles let in around the base.

The details of the invention were given in a paper to the British Association in 1856 under the title, *The Manufacture of Malleable Iron and Steel without Fuel*. The original experiments were carried out with pig iron from Blaenavon purchased from a London founder at a time when Bessemer did not appreciate the considerable variation in the composition of pig iron from different sources. After costly failures among his British licensees arising from the use of medium and high phosphorus irons made from local ores, Bessemer turned his attention to low phosphorus Swedish pig iron initially in an attempt to resolve the question of whether the phosphorus content was the cause of the trouble. In this choice he was fortunate, for these irons also contained sufficient manganese to effect the removal of the oxygen introduced during the blow and thereby to render the steel workable. This important role of manganese had been recognized by Musket in 1856, though he has seldom received due credit for this essential discovery.

In these early years Bessemer found great difficulty in achieving consistent quality, a necessary prerequisite if the process was to be widely used. The unwillingness of the British ironmasters to adopt it led Bessemer to erect his own steelworks in Sheffield and it was here in 1860 under a separate patent that a tilting unit similar to a modern converter was first used.[1] The ultimate development of the Bessemer process for the production of steels of consistent quality was, however, due not to Bessemer and the British steelmakers, but to Göransson in Sweden who introduced a number of improvements in the process and brought to bear the expertise necessary to produce consistent qualities on a regular production basis. The suitability of Swedish pig iron and this early technical contribution from that country are factors which contributed to the common equation of Swedish steel with a high quality product.

Swedish pig iron was extremely low in phosphorus and sulphur, but high in silicon and was commonly called acid pig; the lining used in the converter was an acid material, ganister or silica brick, and the slag formed from the oxidation products was also acid in character. While substantial deposits of ores similar to that from which Swedish pig iron was made existed in the United States and in some parts of Europe, British home ores available at that time were, in general, too high in phosphorus. One result was a considerable British investment in the rich haematite (low phosphorus) ores of Spain in order to secure supplies of material from which a pig iron suitable for use in the acid Bessemer process could be made.

The Bessemer process also suffered from the disadvantage that it could not utilize mild steel scrap which necessarily accumulated in any steel works and in a steel-using, industrial community. Reprocessing of steel scrap had cost advantages, provided an effective, economic means could be found for remelting it. The principle which ultimately led to success—regenerative heating—was first intro-

[1] *J. Iron & Steel Inst. 183* (1956) 179-207.

duced not in the area of steelmaking, but into blast furnace practice in Scotland by the Reverend Robert Stirling and his brother and was incorporated in a neat fashion in the Cowper blast furnace stove. The hot exhaust gases and the air to be fed to the furnace were passed time and time about through brick filled chambers, thereby giving a preheated air blast. Applied to a coal fire, this preheated combustion air considerably enhanced the temperature that could be achieved.

This idea was taken up by Charles William Siemens, a German by birth, who had originally come to Britain to dispose of inventions of his family's electrical business in Germany. With his brother, Frederick, he conceived the idea of using a gas fuel and of preheating separately by regenerative means both the fuel and the combustion air. In its simplest form, four chambers filled with refractory bricks were provided, two at each end of the furnace. At any time, two of the chambers conveyed the hot exhaust gases from the furnace and were thereby heated to a high temperature; the gas fuel and air were passed separately through the other two. After the chambers conveying the exhaust gases were well heated, the flows were reversed, the exhaust gases passing out through the cooler pair of chambers and the inlet fuel gas and air entering through the pair of heated chambers. Continuous repetition of this operation enabled temperatures of 1,100–1,200°C to be achieved in the regenerators, so that when the air and gas fuel were mixed on emerging from the inlet ports, they were already at a high temperature. The result was a fierce combustion resulting in a flame temperature in the furnace sufficient to melt steel. The first Siemens steel plant incorporating this principle began operations in Landore, South Wales, in 1869.

Steel scrap in those years was a highly variable material. The Martin brothers, Pierre and Emile, operating a Siemens' furnace in France under licence were soon to show that, by using a mixture of scrap, wrought iron and pig iron and by working up a lively boil in the furnace with the evolved gases, an improved product could more consistently be obtained. Siemens, himself, had also discovered that, in place of scrap, pig iron could be used and the oxidative refining hastened by adding a quantity of high quality iron ore as a source of oxygen. From these several developments the Siemens-Martin or open hearth process emerged as an alternative to the Bessemer unit for steelmaking. Its particular initial advantage was that it could deal with almost any pig iron-scrap combination. If the mixture were too low in carbon, more of this element could be added in the form of coke or anthracite; if it were too high in carbon and other impurities, iron ore was the appropriate oxidizing additive to facilitate their removal.

Neither the Bessemer converter nor the Siemens-Martin furnace lined with silica was able to handle iron which was high in phosphorus, irrespective of how long the blow or boil was prolonged. By the 1870's, this problem had acquired a degree of acuteness in Britain where steelmakers had become heavily dependent on iron made from low phosphorus haematite ores imported chiefly from Spain and not readily or widely available in Britain. This problem was solved by a London police court clerk, Sidney Gilchrist Thomas, who had acquired a considerable knowledge of metallurgy and applied chemistry from courses he had taken at the Royal School of Mines. With his cousin, Percy Gilchrist Thomas, he recognized that the solution lay in changing the furnace or converter lining from acid silica to a basic refractory such as limestone or dolomite. Such a lined vessel would hold a basic slag capable of extracting and fixing the acidic phosphorus oxides. The apparently simpler solution of adding lime or limestone to the metal charge in an acid lined vessel would not do, since this practice would quickly destroy the silica lining. This discovery by Thomas and the reasoning which underlay it was remarkably simple, indeed so simple that it excited little interest when it was announced to a meeting of the Iron and Steel Institute in 1878. By April 1879, however, successful large scale experiments had been carried out at Bolckow Vaughan's works at Middlesbrough following earlier trials at Blaenavon works in Monmouthshire.

Thomas' modification of Bessemer's process did not, however, meet the particular conditions in Britain where iron produced from local ore was of moderate phosphorus content. Its real success came in Germany where there existed a bountiful supply of hitherto intractable high phosphorus ores from Lorraine. The Thomas process, as the Germans called it, was to them what the acid Bessemer process had been to the North Americans operating on their local, good haematite, low phosphorus ores. In Britain, Thomas steelmaking steadily declined until by 1920 it had completely disappeared. It was revived at Corby in Northamptonshire in 1935, by which time a technique for utilizing iron from the medium phosphorus ores which occur at the bassett of the Jurassic limestone had been worked out. The other factor which influenced its revival at that time was the imposition of tariffs to keep out of Britain imports of tube making skelp hitherto obtained from Germany.

The adoption of basic refractory linings in open hearth furnaces widened considerably the range of raw materials that could be conveniently refined by this method. By 1918, acid and basic steelmaking in Britain were in rough balance, but thereafter basic practice moved steadily ahead, so that by the 1930's basic open hearth practice was the dominant method of steelmaking in Britain.

The three key figures, Bessemer, Siemens and Thomas reveal some remarkable contrasts. Bessemer was a professional inventor and a man of business; Siemens was a scientist and technologist of standing in his own right, as his election to Fellowship of the Royal Society in 1862 reveals; Thomas was a rather mundane clerk who in his spare time had equipped himself in science by attending evening classes, but who, like Siemens, had had little or no previous contact with the iron and steel industry. In each case, they set out to solve problems which were definable; in each instance, the solutions found were essentially simple ones. From a scientific viewpoint, Thomas' contribution had its origin in the recognition of the problem in fairly precise chemical terms; the other two had more direct origins in the preceding industrial practice.

The contributions of Bessemer, Thomas, and the Siemens and Martin brothers led to four established methods of steelmaking, acid and basic Bessemer and acid and basic open hearth. Each was capable of handling certain types of feed, the basic open hearth being the one that was least restrictive. In addition to these thermal processes, the progressive increase in availability and decrease in cost of electrical power as a source of energy led to the introduction and extension of electric steelmaking. Generally speaking, this was concerned primarily with the production of special steels mainly from selected scrap on a unit scale considerably smaller than those for conventional processes. It did, however, offer a possible means of making common steels from scrap that was not conveniently employed in low scrap-using, conventional processes of the converter type.

In the following paragraphs, the essential technical features of conventional steelmaking processes are briefly reviewed, not with any aim of completeness or of sophistication, but to provide a base from which the development of oxygen steelmaking processes may be appreciated.

BESSEMER PROCESSES

The typical situation in the acid Bessemer process is that air is blown through molten pig iron containing carbon, silicon and manganese in a converter lined with silicious material. Carbon is removed as its volatile oxides and the other elements form a ferrous-manganese silicate slag. The oxidation of these elements produces heat which not only maintains the charge at the desired temperature, but may also increase it, especially if silicon is present in large amounts. In such cases, some cold scrap may be added to control the temperature. The initial product is usually a very mild steel containing considerable amounts of oxygen and nitrogen, to which spiegeleisen or ferromanganese is added before casting to yield steel of the required composition and properties. Control of the process is usually in terms of the nature of the flame emitted from the converter. Initially, when silicon and manganese are being oxidized, the flame is short and of low luminosity which increases in length and luminosity during carbon elimination and then greatly decreases. Blowing is then discontinued, otherwise brown fumes, indicating the oxidation of iron and hence the loss of this element, are given off.

In the basic process, the pig iron typically used is low in silicon and manganese, but high in phosphorus. The vessel is lined with a basic refractory and lime or limestone is added to the converter before introducing the pig iron. During blowing, carbon is eliminated first, but in this case the blowing is continued after the flame drops—the so-called afterblow—to eliminate the phosphorus. The principal components in the slag are calcium phosphates and from a high phosphorus iron the slag containing 18–22 per cent phosphorus pentoxide (P_2O_5) may be disposed of as a fertilizer.

The Bessemer converters are essentially refractory lined, pear shaped vessels often made in three sections to facilitate relining. Blowing is achieved with air at a pressure of 10–30 lb per square inch through a set of tuyers let into the base of the converter which is mounted on trunnions giving rotation through about 200° to facilitate charging, blowing and emptying. Units vary considerably in size, but are usually from 10 to 60 tons capacity. Those operating in Britain in 1963 are listed in Table 4. The time taken for an average blow is about twenty minutes, so that small plants are capable of high rates of production; for example, a plant comprising three 25 ton converters is capable of making 700,000 to 800,000 tons of steel per annum.

Table 4: Bessemer converters in United Kingdom, 1963[1]

Type	Location	Number	Capacity (Tons)
Acid	Workington Iron & Steel Co., Workington	2	25
Basic	Stewarts and Lloyds Limited, Corby	5	28
Basic	Steel Company of Wales, Abbey	4	60

The air requirements will, of course, vary with the amount and nature of the impurities to be oxidized, but will generally be in the range 5 to 8 tons per ton of iron blown. The yield of sound metal in relation to pig iron fed depends a good deal on the degree of process control achieved, particularly of the ejection of metal from the mouth of the converter and, in the basic process, of the afterblow. Figures of 87–89 per cent for acid practice and slightly lower figures for the basic process are commonly quoted. In the basic process, lime additions will depend upon the phosphorus, silicon and manganese contents, but may amount to 280–330 lb of lime per ton of metal.

The distributions in the thermal balances are summarized in Table 5.[2]

Table 5: Thermal balances in acid and basic Bessemer practice

Input	Acid %	Basic %	Output	Acid %	Basic %
Molten metal	33	38	Molten steel	46	42
Slag formation	7	8	Molten slag	10	20
Oxidation of			Exit gases	24	25
C,Si,Mn, and Fe.	60	54	Losses	20	13

The hot metal supply is obtained by melting pig iron in a separate cupola, by direct feeding from a blast furnace, or more usually in modern practice from a mixer. This is essentially a reservoir of molten pig iron between the blast furnace and the converter with a capacity large enough to even out some of the variations in composition in the supply of pig iron. Since sulphur removal takes place erratically during the blow, it is a common procedure, where this element is abnormally high in pig iron used in basic practice, to carry out desulphurization by the addition of soda ash to the mixer or transfer ladle before feeding to the converter.

The making of Bessemer steels depends upon reducing by air oxidation and slagging the concentrations of unwanted impurities to

[1] Annual Statistics 1963, Iron & Steel Board and British Iron & Steel Federation, pp. 46, 48.

[2] Adapted from G. R. Bashforth, The Manufacture of Iron & Steel, 2nd ed., 1964, pp. 104, 124.

levels required in the finished steel. This usually also results in the reduction of the desirable impurities of manganese, and in some cases silicon, below the levels required. Additions of spiegeleisen, ferromanganese or ferrosilicon to meet a final specification are therefore commonly needed. These additions serve the dual purposes (i) of 'killing' the steel, that is of removing to the slag by the addition of the more reactive manganese and silicon the oxygen introduced during the blow, and (ii) of subsequently adjusting the manganese and silicon contents of the fully killed steel to the desired levels. Made under careful control, both types of Bessemer steel are suitable for many applications. The major disadvantage for applications involving deep drawing and pressing arises from the high nitrogen content, the effect of which is aggravated by high phosphorus contents. A comparison of nitrogen contents of blast furnace metal, Bessemer steels and open hearth steels is given in Table 6.

Table 6: Nitrogen contents of blast furnace metal and steels

Material	Nitrogen content %
Blast furnace metal	0·002–0·006
Bessemer steels	0·010–0·020
Open hearth steels	0·004–0·007

A general disadvantage of the Bessemer processes is that they consume negligible quantities of scrap and will accept only a hot metal charge. They do not, however, require additional fuel.

OPEN HEARTH PROCESSES

In the Bessemer processes, the slag formed acts as a convenient receptacle for the oxidation products other than the oxides of carbon. In open hearth processes the slag performs the dual roles of providing the oxidizing medium and of receiving and retaining the oxidation products. The oxygen is provided to some extent by the furnace atmosphere, but predominantly by the addition of iron ore or mill scale or from rust present on the scrap incorporated in the charge. Unlike the Bessemer processes, there is a source of heat separate from that generated in the refining operation.

Open hearth furnaces are essentially simple in design and construction. In addition to the regenerative chambers for preheating the fuel and combustion air, the valve system and fume exhaust, the furnace itself comprises a working hearth, back and front walls and roof. The hearth, constructed of acid silicious or basic dolomite or magnesite refractory, is contained within a heavy reinforced steel pan and is finished with a slope to a tap hole to facilitate the removal of molten metal and slag. Open hearth furnaces may be of either the fixed or tilting types. In the latter the hearth may be tilted about 10° to the charging side for slagging and 20–30° for tapping, thereby facilitating multiple slagging practice and, as a result, extending the range of materials that may be used as a feedstock.

The acid open hearth process is unable by its nature to remove impurities of phosphorus and sulphur, and its feed needs, therefore, to be low in these elements, usually less than 0·05 per cent phosphorus and 0·040 per cent sulphur. Silicon is not required as a fuel, but there must be sufficient present to give an adequate slag volume, 1·2 to 2·7 per cent with an average around 2·4 per cent being fairly common. While the manganese content is not critical, too high a level of this element will retard the oxidation of carbon, whereas inadequate quantities tend to lead to overoxidation. The acid process may be divided into five stages: charging; melting; the melt out; the boil; finishing. After charging with pig iron and scrap, complete melting is achieved usually in the temperature range, 1,520–1,570°C. Some silicon and manganese will be removed as a slag with the existing iron oxide at this stage. After additional heating, the iron ore feed is begun and, during this melt out stage, additional silicon and manganese are removed to the slag which becomes richer in silicon. When the silicon in the metal has fallen to 0·15 per cent, the carbon boil commences and is maintained by careful control of temperature and ore additions and, if necessary, by the addition of some limestone to facilitate the metal-slag reactions. The finishing stage which prepares the charge for tapping may follow one of several alternatives, but all are designed (i) to obtain the lowest possible iron content in the slag consistent with fluidity, (ii) to achieve a stable equilibrium between the metal and slag to avoid uncontrolled reactions in ladle or mould, and (iii) to minimize non-metallic inclusions by achieving the correct slag viscosity and activity, or by the judicious use of deoxidizers. Alloy or other additions required to adjust the final metal composition are made before tapping.

The brief outline given above serves to illustrate that the critically important feature in making acid open hearth steel is the control of the slag, both as regards its chemical composition and physical state. If properly made, these steels are fairly free of inclusions, low in oxygen and nitrogen and are suitable for high grade applications. Alloy additions can conveniently be made under controlled conditions so that the process has advantages in the production of alloy steels.

The basic open hearth process will accept a wider range of raw materials than any other conventional steelmaking process. Sulphur and phosphorus may be eliminated, and the only restriction is that the initial silicon content should not be excessive. This latter requirement arises because of the need to avoid the formation of too thick a slag layer which would interfere with the heat transfer between the gas above and metal below, and because of the possible incidence of excessive attack on the basic refractory linings. As in the acid process, the slag serves as the receptacle for oxidized impurities and as the carrier of the oxidant. Carbon is, as before, removed as the volatile oxides and the slag has to cope with silicon, phosphorus, manganese and sulphur. To remove these elements effectively, a balance has to be struck between three properties, the oxidative character of the slag, its basicity and the maintenance of the appropriate physical properties to facilitate the slag-metal reactions. As changes in the composition of the slag are brought about by the addition of iron ore and lime, impurities are progressively taken up thus causing the properties of the slag to undergo further change. The important point here is that the oxidation of impurities, their removal to and retention by the slag are not to be looked upon as a series of independent processes, but as a group of highly interdependent events. Consequently, a number of local practices has evolved which depend a good deal on the experience of the operators assisted by a number of simple, or not so simple, empirical tests of the properties and composition of the slag during the several stages of the refining operation. It is sometimes convenient and profitable to adopt a two slag practice, especially where the initial charge is unduly high in phosphorus; this generally results in improved control at the later stages of refining.

Overriding all these matters, there remains the question of the control of temperature. As one major heat input term, namely that arising from the combustion of the fuel, is capable of control independent of the heat generated by oxidative reactions in the furnace charge, open hearth steelmaking is thermally flexible and sophisticated in comparison with Bessemer techniques. It also lends itself to increases in scale to 300 and even 500 tons. In Britain as in most parts of the steelmaking world, open hearth furnaces have grown substantially in unit size since the Second World War, the largest currently operated in the United Kingdom being four 400 ton basic furnaces at the Margam works of the Steel Company of Wales.

The metallic yield in proportion to the charge varies somewhat depending on the nature of the latter. With a charge comprising hot pig iron with substantial additions of iron oxide, the yield may exceed 100 per cent, this situation arising from the iron produced from the added oxide. With a large proportion of scrap in the charge, yields will usually be of the order of 92–94 per cent, and with poor grade pig iron they may fall to 87–88 per cent. Tap to tap time will also depend on the composition of the charge and on whether this is hot or cold metal. A round figure of ten to twelve hours for conventional practice will serve adequately for purposes of rough comparison of output rates.

The preponderance of open hearth steelmaking methods in the United Kingdom in the period 1930 to 1955 is illustrated in Table 7.[1]

Table 7: Production of crude steel by process in United Kingdom, 1930–55
(Thousand tons)

| Year | Bessemer | | | Open Hearth | | |
	Acid	Basic	Total	Acid	Basic	Total
1930	279	—	279	1,805	5,099	6,904
1935	199	224	423	1,857	7,361	9,218
1940	176	738	914	2,174	9,274	11,448
1945	171	687	858	1,159	9,026	10,185
1950	248	846	1,094	1,311	12,981	14,292
1955	252	1,032	1,284	1,000	16,252	17,252

There were many reasons for this situation. Probably the most important were the sustained availability of scrap at prices generally below those for pig iron and the great flexibility in the proportions of pig iron and scrap that could be used in the charge. Moreover, home deposits of iron ore in Britain are not, in general, specially suitable for making a pig iron that is ideal for Bessemer steelmaking. The usages of lime, ore, mill scale or finishing alloys are not sufficiently different to be significant, but against the advantages of the open hearth has to be set an added fuel cost for about 40 to 65 therms per ton depending on the extent of hot metal in the charge.

A further consideration concerns the life and replacement of refractory linings, much of which has to be carried out by hand labour. It is usual to operate Bessemer converters in pairs, so that one unit

[1] *Iron and Steel Annual Statistics, 1963.* Iron and Steel Board and British Iron and Steel Federation, p. 53.

can be relined while the whole of the rest of the plant operates with the alternative converter. With open hearth furnaces, being larger in size and cost, the rebuilding and relining operations may be more protracted tasks and result in immobilizing a larger fraction of other capital facilities. Against this, must be set the fact that the life of open hearth linings is usually larger both in terms of time and tonnage of metal produced.

The questions of production rates and capital costs are often rather vexed ones. Taken in isolation, a typical, large 300 ton open hearth furnace working on a tap to tap time of twelve hours— nominally twenty-five tons per hour—will just match a small Bessemer converter of twenty-five tons operating with a blowing time of twenty minutes and a total turn round cycle of, say, one hour. There is little argument that the capital cost of the open hearth is considerably higher. However, steelmaking seldom exists in isolation and the required production rate may be determined by those processes which precede and follow it and with which it may have to keep in step and in style. Open hearth steelmaking having once been established on an extensive scale within a given works or on a national scene, there are considerable technical, economic and social pressures to sustain it. On the technical side, flexibility with respect to feed and to product usage are dominant; on economic grounds, the regenerated investment arising from rebuilding and relining is one which most companies are very loath to scrap and write off, especially in times of uncertain demand; on the social side, superintendents and crews tend to develop considerable judgment, skill and even artistry which, perhaps more than in many other fields of technological endeavour, provide important sustenance and create an intangible asset which managements are often rather unwilling to discount.

ELECTRIC STEELMAKING

Electric furnaces have quite a long history in the iron and steel industry, but until quite recent years were employed almost exclusively in the production of steel castings and in the manufacture of alloy steels. We are not concerned with this area in the present study, and the discussion will be wholly restricted to the production of bulk, commercial steels by this means.

In its simplest form, the manufacture of bulk steels in an electric furnace resembles basic open hearth practice, the difference being that the thermal requirements are provided not by fuel combustion, but from electrical energy. The furnace is essentially a refractory lined pot with the hearth and, in modern basic practice, the sides lined with basic refractory. The roof is generally lined with silica and through this electrodes are inserted. The heat is generated by the arcs formed between the electrodes and the charge and by the resistance which the charge offers to the passage of the current. Since the charge is initially cold and solid and finally hot and molten, an important feature is the provision of automatic control of the electrodes, hence the length of arc and thereby the input of heat. The electrodes are normally made of graphite or amorphous carbon and are consumed at rates which may vary from 14 to 30 lb per ton.

One peculiarity of arc furnaces is their load characteristics. During melting down of the charge, violent fluctuations in load occur. In this stage, a high proportion of the peak load is required, but during the later refining operation the load factor may be as low as ten per cent. Since electricity is sold at a price which is dependent on the load factor, it is usually necessary to operate several furnaces staggered with respect to stage in order to maintain a better overall load factor.

Bulk commercial steels made electrically generally follow a simple slag practice. The charge consists of steel scrap, lime or limestone and a carburizer and the reactions and their control follow fairly closely basic open hearth practice adapted to the peculiarities of the charge, the furnace and the different degree of temperature control that is possible. Originally, electric steel furnaces were quite small, ten tons or less in capacity, but these have now grown for commercial bulk steel production to 100 or even 200 tons, and at this level can be considered a real replacement for the basic open hearth furnace employing cold metal practice. The kinds of factors that have made this development possible include the following:

(i) Lower material costs of all scrap charges.

(ii) Lower capital costs and space requirements than for an equivalent open hearth installation.

(iii) Some offset in the differential cost of electric versus fuel heating by the much higher thermal efficiency of the former (64 per cent as against 29 per cent).

(iv) Some improvement in flexibility of operation and quality of steel.

The two developments of note in the United Kingdom in this area are the two 60–65 ton direct arc furnaces installed at the Round Oak Steel Works in the Birmingham area where scrap supplies are specially favourable because of the nature of the industries in the region. The second comprises six 110 ton furnaces installed by the Steel, Peec hand Tozer branch of The United Steel Companies at Rotherham. In both cases, these installations were extensions of or replacements for cold metal open hearth furnaces. The general statistics for the United Kingdom set out in Table 8[1] include a small component of 100,000 to 140,000 tons of crude steel made by induction electric furnaces. These furnaces seem unlikely to be used on any significantly larger scale in the manufacture of crude steel.

Table 8: Production of crude steel in United Kingdom by electric furnaces
(Thousand tons)

Year	Production	Year	Production
1930	76	1950	736
1935	107	1955	1,098
1940	435	1960	1,685
1945	542	1963	2,077

An assessment of the significance of electric steelmaking in Britain is afforded by a comparison of the total for 1963, namely 2·1 million tons, against a grand total for all methods of 22·5 million tons. Its importance in our present context rests principally on the fact that it is a major consumer of steel scrap and, as such, may contribute significantly as a balancing component in the introduction of a new technology which may consume less scrap than the conventional open hearth furnace.

The development of tonnage oxygen

For our purposes, air may be regarded as a mixture of 79 mole per cent nitrogen, 21 mole per cent oxygen. The rare gases may for many practical purposes be neglected, but attention has to be directed to the small amounts of carbon dioxide and water which are normally present. These and other impurities such as acetylene which, though present in trace amounts, accumulate in large scale oxygen plants, have to be removed either chemically in advance or by including appropriate measures in the main separation process. Broadly speak-

[1] *Iron and Steel Annual Statistics, 1963.* Iron and Steel Board and British Iron and Steel Federation, p. 53.

ing, there are two ways in which the separation of oxygen and nitrogen may be achieved. The first is a chemical approach in which the oxygen is selectively combined with a reagent and subsequently regenerated in the pure state. The second makes use of the differences in physical properties between oxygen and nitrogen. Except on a small scale, the second alternative is overwhelmingly the preferred approach, to which our present discussion is exclusively directed.

The theoretical principles underlying the physical separation of oxygen and nitrogen from air were fairly well understood by the close of the nineteenth century, but the development of continuously operating, stationary state processes based on these principles is essentially an achievement of the twentieth century. The thermodynamic properties of oxygen-nitrogen mixtures require that the separation at any temperature involves two basic steps, liquefaction by means of compression and refrigeration, and distillation of the resulting liquid. If the products leave the separation plant at room temperature, the net refrigeration required after the initial cooling down period will comprise that needed to offset the ingress of heat from the surroundings and a component arising from the imperfections in the heat exchangers employed. If, on the other hand, cold liquid products are required, a great deal more refrigeration has to be supplied. Without considering any formal process, it is obviously cheaper on energetic grounds to supply the separated components at room temperature if this is the condition of the intake air. The supply to individual works of oxygen or nitrogen in the form of their respective liquids is an intrinsically expensive method for large tonnages.

A second important consideration is whether either or both gases are required at the same location and at what levels of purity. In a very sophisticated situation, both may be required at high purity (99·5 per cent), but this is likely to be rare. In other cases, only one of the gases is required in this state, the other being vented or used comparatively inconsequentially. Another situation of practical significance is that in which one or possibly both gases are required at two levels of purity, say 99·5 per cent and 95 per cent. These varying requirements lead to differences in detailed design, in operational characteristics and in capital cost. These more detailed requirements may be looked on as evolutionary steps from a central, original development.

In its rudimentary form, the separation of air into oxygen and nitrogen involves the distillation of liquefied air through a rectifying column, but, before such a proposal can have any hope of commercial exploitation, it is essential that the processes of liquefaction and distillation proceed from intake to output continuously in a steady manner. To do this, it is necessary to establish a cyclic operation which is balanced with respect both to heat and mass. The fundamental principle conceived and utilized by Linde and Claude was that the air to be separated could itself be used to provide the required refrigeration. This is achieved by compressing the incoming air to a suitable pressure and subsequently expanding it either through a valve, thereby making use of the Joule-Thomson effect, or in an expansion engine or turbine in which the air is made to do work. The simplest arrangement, shown in Fig. 1, illustrates the original Linde cycle first operated in 1902.

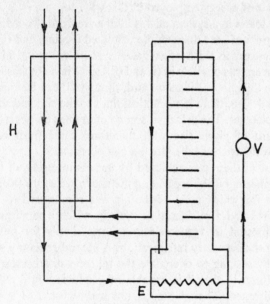

FIG. 1 Single column Linde cycle

Compressed air, after the removal of water and carbon dioxide, is introduced through the heat exchanger H which is cooled by the returning cold products, and traverses a cooling coil in the sump E before being expanded to atmospheric pressure through the Joule-Thomson valve V. It is then delivered mainly as liquid to the top of the distillation column. The descending liquid is progressively enriched with oxygen until, in the stationary state, impure nitrogen is continuously withdrawn from the top of the column and liquid or gaseous oxygen is continuously withdrawn from the bottom. To enable liquid oxygen to be withdrawn in such a plant, pressures up to 200 atmospheres are required to provide sufficient refrigeration, but if gaseous oxygen at atmospheric pressure is required, compression pressures of 60 atmospheres will suffice. The energy advantage of operating an on-site plant producing gaseous oxygen is represented by this difference in pressure. As the energy expended in compressing the incoming gas is the major operating cost of all air separation plants, and usually represents more than half the total cost of delivered oxygen when due allowance is made for capital charges, labour and ancillary services, special attention is directed to reducing the amount of energy employed in this way.

Energy losses in a system of this kind arise from two main causes: (i) the ingress of heat through the insulated lagging, and (ii) the imperfect recovery in the heat exchanger. There are practical limits to the improvements possible in (i) and (ii), since it is not feasible to use an infinite thickness of insulant and, though (ii) can be improved by advanced design, there is a practical limit to the size and the cost of a heat exchanger. The principal source of savings on operating costs for a plant of given size will, therefore, come from economies effected in the energy expended on compression.

The first of these was achieved by not compressing all the air to 60 atmospheres, but by separating the single cycle into two, the process cycle for which most of the air was compressed to only 5–6 atmospheres and a refrigeration cycle in which a small proportion was compressed to 150–200 atmospheres simply for purposes of providing the necessary refrigeration. A second economy was made possible by seeking to overcome the inherent disadvantages of the single column in that it is capable of producing pure oxygen and impure nitrogen. Though this is not a drawback, as such, if the nitrogen has little or no commercial value, substantial amounts of oxygen are, in effect, being wasted. For example, a single column yielding 100 per cent oxygen will give a top column gas comprising 93 per cent nitrogen and 7 per cent oxygen. The energy used in processing this 7 per cent oxygen is not recovered.

This disadvantage may be overcome by employing a double column which was evolved by Linde in 1910 and still predominates today. It is illustrated in Fig. 2. The compressed air, precooled in the heat exchanger, enters the evaporator coil P and is expanded to 5 atmospheres through the valve V and delivered to the middle of

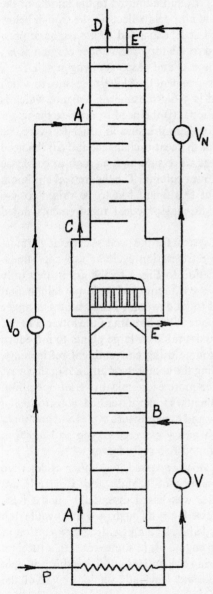

FIG. 2 Double column
Linde cycle

the lower column at B. Here it is separated into pure liquid nitrogen and a liquid containing about 38 per cent oxygen. The latter, withdrawn from the bottom of the lower column at A, is expanded to atmospheric pressure through V_O and admitted to the middle of the upper column at A'. The liquid nitrogen collects in the trough below the condenser and is removed at E, expanded to atmospheric pressure through V_N and admitted to the top of the upper column at E'. Gaseous oxygen is withdrawn at C and gaseous nitrogen at D.

We may note that in this arrangement the highest pressure to which the process air is compressed is 5–6 atmospheres, a figure which is determined by two considerations: (i) the need to operate the upper column slightly above 1 atmosphere pressure in order to overcome the resistance of the plates and the heat exchanger; and (ii) the need to operate the lower column at a pressure high enough to condense the nitrogen at the top of the lower column. The theoretical minimum value is 3·6 atmospheres, but this figure has to be raised to 5–6 atmospheres to meet the additional practical requirements noted above.

If, now, the double column system for the process cycle at 5 atmospheres is combined with the refrigeration cycle at 150–200 atmospheres, there is some modification to Fig. 2 to the extent that only the refrigeration cycle air traverses the sump of the lower column and the process air is now cooled to its dewpoint in the heat exchanger and admitted as saturated vapour at the bottom of the lower column. In addition, it may also prove desirable on large plants to introduce some external refrigeration source using conventional refrigerants, such as ammonia, for precooling the compressed air and, in this way, achieve a more flexible and/or more economical overall operation. To summarize, the developments so far described constitute the Linde double column with low and high pressure cycles and ammonia refrigeration from which high purity gaseous oxygen and nitrogen may simultaneously and continuously be withdrawn.

Shortly after the Linde liquefier appeared, two other cycles were developed with which the names of Claude and Heylandt are associated. The Linde liquefier was based essentially on the isenthalpic expansion of compressed air through a valve, while the alternative pioneered by Claude involved an isentropic expansion in the cylinder of a reciprocating engine or, in some cases, in a turbine. There were, however, engineering difficulties in utilizing this principle alone and in both the Claude and Heylandt versions, part of the compressed air was expanded through a valve and part in an expansion engine. The important point is that the use of the Claude-Heylandt system enables the refrigeration performance to be improved, a feature which has proved valuable where the load is very high, as in plants required to deliver liquid oxygen and in small scale and mobile units. Expansion turbines, as distinct from reciprocating engines, have the additional advantage that for low pressure differentials they may be designed to operate with comparatively high efficiency.

A further contribution due to Claude was that of dephlegmation which, for some product patterns, has been incorporated in some recent air separation plants. In condenser-evaporators in which the vapour condensing within the tubes and the liquid boiling between the tubes are of the same binary mixture, a temperature head tends to be established between the top and bottom of the apparatus inside the tubes. As a result, some difference in composition in respect of each phase between the top and bottom is established. In the case of air, the vapour in the upper part will contain less oxygen than that condensed at the bottom with, consequentially, a stream of comparatively warm gas moving upwards and a stream of cold liquid trickling downwards along the walls of the tubes. The liquid progressively becomes enriched with oxygen as it moves downwards and the gas enriched with nitrogen as it moves upwards. If the condenser is long enough, the vapour emerging at the top may be almost pure nitrogen and the liquid escaping at the bottom of a composition in equilibrium with the gaseous mixture entering the tubes. This crude form of rectification can be incorporated with economic advantage in some air separation plants.

All the developments noted above were introduced within the framework of comparatively small-scale requirements of oxygen for such activities as oxy-acetylene cutting and welding and for the somewhat larger unit demands for pure nitrogen in the chemical industry for ammonia synthesis. The high and low pressure cycles operating with a double column made it possible to produce high purity oxygen and nitrogen simultaneously, though both were seldom required at the same site and in the relative amounts available in the atmosphere. It was, however, only towards the latter end of the 1930's that these developments reached a commercially exploitable form. At about this time, it began to be evident, especially from work done in Germany, the United States of America and in Russia, that there was a considerable potential use for oxygen in the steel, chemical and gas making industries provided that it could be produced more cheaply than the isolated, rather small scale operations had hitherto permitted. This led to renewed efforts to seek points at which further cost reductions might be made.

The pressures in the two columns of the Linde system are set by thermodynamic as well as operational requirements and cannot be changed sufficiently to yield significant gains. If, however, this pressure difference were utilized to produce additional refrigeration by expanding a certain amount of the nitrogen through a turbine instead of a valve, the amount of air passing through the high pressure cycle and requiring compression to 150–200 atmospheres could be reduced from 17 to 6 per cent with significant savings in power. Another important change was the replacement of the main heat exchanger by pairs of alternating regenerators of a type which had been patented[1] by Fränkl in 1925 and suggested by him in 1928 for air separation plants. In the heat exchangers originally used, incoming warm gas flowed through one set of tubes and was cooled by the cold outgoing gas passing through an independent set. If the gas paths were made very long, reasonably complete exchange was possible, but as the tonnage to be processed grew, so also had the size of the exchanger, if the same efficiency was to be achieved. Ultimately, physical size and concomitant capital cost imposed an upper limit.

The alternative proposed by Fränkl was to replace the heat exchanger by a pair of regenerators which could be used alternately, the 'cold' being stored in the high heat capacity packing of one regenerator during each half cycle. Such regenerators were comparatively small, cheap, and easy to manufacture, they provided a large surface area per unit of volume and they offered only a low resistance to gas flow. Since, however, they were used alternately, it was difficult to obtain highly pure products unless at the beginning of each half cycle a certain volume of the alternate gas was first purged, thereby reducing somewhat the yield. If this practice was not followed, gas with a purity of 98 per cent only was obtained. A further improvement came from replacing the earlier reciprocating compressors with more efficient turbocompressors.

With these various improvements, Linde-Fränkl plants in Germany had reached a high degree of refinement by the beginning of the Second World War. After 1946, numerous other plants began to

[1] British Patent 335, 257.

make their appearance on the industrial scene. These incorporated sundry, additional modifications and improvements, including reversing exchangers of the Collins type which obviated the contamination problem of Fränkl regenerators, modified cycles, the use of improved materials of construction and extensive systems of automatic control. Moreover, the scale of unit production increased considerably giving rise to the term, 'tonnage oxygen'. The nomenclature varies somewhat in different countries, but, in general, tonnage oxygen refers to oxygen made at high purity (99·5 per cent) or medium purity (95 per cent) or both in plants located within or adjacent to the users' works and operating on a scale greater than 50 (in some countries 30) tons per day, the product being delivered to the point of use at the required pressure.

This situation did not arise solely to meet a need in the steel industry, but in the first instance to satisfy a requirement of the chemical, gas and petroleum industries. By 1961, the use of tonnage oxygen for steelmaking had, however, grown rapidly as the data[1] in Table 9 illustrate. The growth rate at that time was estimated to be between 10 and 20 per cent.

Table 9: Industrial utilization of tonnage oxygen
(Capacity in tons per day)

Country	For chemicals	For steelmaking and other uses	Total
U.S.A.	8,415	13,800	22,215
W. Germany	4,220	4,045	8,265
Italy	4,475	305	4,780
U.K.	1,200	1,900	3,100
France	1,040	1,460	2,500
Spain	1,185	160	1,345
Netherlands	1,050	380	1,430
Belgium	660	1,470	2,130
Canada	495	805	1,300
S. America	180	265	445
Others (excluding Soviet Bloc)	4,335	3,510	7,865
Total (excluding Soviet Bloc)	27,275	28,100	55,375
Soviet Bloc (estimated)			17,000
World Total			72,375

Among the companies supplying tonnage oxygen plants, the following appear to command most of the market:

U.S.A.   Linde Co.; Air Products; General Dynamics; National Cylinder Gas; Hydrocarbon Research.

U.K.   British Oxygen Company; Air Products.

France   L'Air Liquide (France).

[1] M. Sittig, Chemical and Engineering News, 39, 92, 27 November 1961.

Germany   Linde (Gesellschaft für Lindes Eismachinen A.G.);
Adolf Messer GmbH.

The sizes of individual plants within the tonnage range vary from the lower limit of 50 tons per day to as much as 1,000 tons per day. As early as 1949, Hydrocarbon Research had installed two 1,000 ton per day units in Brownsville, Texas, but, in general, units tend to be in the range 100–500 tons per day. The supply of tonnage oxygen for chemical or steelmaking processes is usually arranged in one of two ways: (i) on an over the fence basis from a plant built, controlled and operated by the supply company adjacent to the customer's works; or (ii) from an oxygen plant built as an integrated part of the chemical or steel works by the supply company, but owned and operated by the oxygen user.

In the United Kingdom, the first method is the predominant one with the British Oxygen Company controlling most of the market. In 1965 production of oxygen was running at a level of 4,500 tons per day from this company's 26 plants.[1] Three additional plants came into operation in 1966 resulting in an increase in output of 20 per cent. For large consumers, gaseous oxygen is supplied at a tariff of about £5 per ton,[2] a figure which increases rapidly for small users. Costs and hence tariffs vary considerably according to unit size. In the United States, for example, it has been estimated that for a 100 ton per day plant the cost runs at $16–18 per ton and at 1,000 tons per day it has been suggested that the figure could fall to about $8 per ton. Data[3] computed by the U.S. Bureau of Mines in 1960 are set out in detail in Table 10.

Table 10: Operating cost and selling price of 99·5 per cent oxygen
For: Split cycle operation; 450 p.s.i.g. discharge pressure;
power allowed at $0·0075 per unit

Size tons per day	Operating cost $ per ton	Selling price per ton		
		12·5%	20%	30%
		return or investment		
150	10·00	15·14	18·57	22·86
200	8·98	13·44	16·41	20·12
500	6·92	9·97	12·01	14·54
1,000	5·95	8·25	9·78	11·69

[1] British Oxygen Company—*Sunday Times*, 9 May 1965.
[2] *Engineering 196*, 236 (23 August 1963).
[3] S. Katell and J. H. Faber, Bureau of Mines Circular 7939, U.S. Dept. of Interior (1960).

Bashforth[1] has quoted some 1959 figures due originally to Harrison in the following terms.

Output tons per day	50	125	300
Cost per ton (£)	5·76	4·16	3·22

The exact figure here is not critical, but two points need to be noted. The first is that the British Oxygen Company is able to quote tariff rates competitive with those ruling abroad and it can scarcely be maintained that the existence of a virtual monopoly held by this company in Britain has deterred the British steel industry from introducing oxygen steelmaking processes. In more recent years, a useful element of competition has been introduced by Air Products. The second point concerns the contribution to the cost of steel arising from the intensive use of oxygen. For large installations of, say, 200 tons per day, a representative tariff would not be greater than £4 per ton; intensive oxygen steelmaking in a converter currently utilizes, at most, 2,000 cu.ft. oxygen per ingot ton costing about 6/-. If an average factor of 1·36 is allowed for the conversion of ingot to finished product tons, the oxygen cost is 8·2/- per ton of finished product for which the selling price is, say, £34 per ton. This calculation shows that the cost contribution arising from the use of oxygen in steelmaking is small. Moreover, a plant making one million tons of crude steel a year by intensive oxygen practice will require oxygen supplies of the order of 200 tons per day. At this scale, the further savings possible by enhanced size of the oxygen plant with co-ordinated distribution to several steel plants in a locality will be modest and may be offset by the increased distribution costs incurred.

One of the problems in the supply of tonnage oxygen for steel-making arises from the fact that the oxygen producing unit is, by its nature, geared to steady operation, whereas the steelmaking units themselves make variable demands. To meet this situation, it is necessary either to make a certain proportion of the oxygen as liquid for storage (at considerably higher cost), or to vent surplus oxygen at times of minimum demand, or to attempt to balance the demand by some kind of grid supply to several consumers. In Britain, the last named practice has been instituted by the British Oxygen Company in Sheffield where there exist within a limited area both large and small consumers. In the future at least, the question of oxygen

[1] G. R. Bashforth, *The Manufacture of Iron and Steel*, vol. II, 3rd edn., Chapman & Hall, London, 1964, p. 169.

supplies will be a factor to be considered in plant location, but this will probably do no more than underline an already existing situation arising from the supply of hot metal.

The use of oxygen by the United Kingdom iron and steel industry in recent years is set out in Table 11,[1] in which the figures exclude the consumption by re-rollers and iron founders. It is instructive to classify these uses into three categories as follows:

(i) iron making—item 1

(ii) steelmaking—items 2 to 5 inclusive

(iii) miscellaneous—items 6 to 11 inclusive

*Table 11: Oxygen consumption in U.K. Iron and Steel Industry
million cubic feet (60°F and 30 in Hg.)*

Item No.	Plant where used	Volume used					
		1958	*1959*	*1960*	*1961*	*1962*	*1963*
1	Blast furnaces	136·8	191·5	248·7	268·8	2,347·0	4,888·5
2	Pre-refining for steel furnaces	174·5	248·6	297·1	288·3	194·0	184·4
3	Open hearth furnaces Air enrichment (combustion)	896·9	1,726·1	1,896·8	1,820·8	2,552·3	2,565·6
	Bath injection (refining)	530·3	830·8	1,362·8	1,823·1	3,568·1	4,000·7
4	Electric furnaces	194·4	216·5	281·7	282·9	255·8	422·1
5	Converters	80·9	604·4	1,991·0	2,572·6	3,479·5	6,260·9
6	Miscellaneous melting shop uses	—	310·1	396·5	369·5	378·1	529·1
7	Scarfing, dressing and cutting of ingots and semis	1,295·2	1,368·0	1,510·1	1,672·5	1,851·5	1,934·2
8	Steel foundries		220·0	261·6	311·7	282·7	239·1
9	General engineering	616·8	469·1	559·5	593·9	615·9	515·2
10	Scrap preparation	745·7	283·8	301·8	322·4	300·5	311·4
11	Other uses		394·8	569·5	588·7	1,142·7	1,617·1
	Total	4,671·5	6,863·7	9,667·1	10,915·2	16,968·1	23,468·3

and to consider the consumption in these areas as percentages of the total in relation to the 1958 figures taken as a base year. The results of this exercise, shown in Table 12, reveal that the steadily increasing use of oxygen in steelmaking preceded that in ironmaking until 1962 when utilization in both areas began to rise quite dramatically.

[1] *Iron and Steel Annual Statistics, 1963*, Iron and Steel Board and British Iron and Steel Federation, p. 21.

Table 12: Distribution of total oxygen consumption, 1958–63

Use	1958		1959		1960	
	%	Index	%	Index	%	Index
Ironmaking	2·9	100	2·8	140	2·6	182
Steelmaking	40·1	100	52·9	193	60·1	310
Miscellaneous	57·0	100	44·3	115	37·3	136
Total	100	100	100	165	100	232

Use	1961		1962		1963	
	%	Index	%	Index	%	Index
Ironmaking	2·5	197	13·8	1,720	20·8	3,570
Steelmaking	62·1	362	59·3	536	57·2	1,395
Miscellaneous	35·4	145	26·9	176	21·9	194
Total	100	262	100	363	100	502

The development of oxygen steelmaking processes

Conventional steelmaking processes were classified in three groups: (i) Bessemer acid and basic pneumatic processes; (ii) open hearth acid and basic processes; (iii) electric steelmaking. In traditional practice, (iii) was associated with remelting and reconstituting steel scrap where the refining reactions compared with (i) and (ii) were minimal. The pertinent difference between (i) and (ii) lay in the role of gaseous oxygen. In the former, no additional heat was supplied, the requirements being met from the direct oxidation of impurities by the oxygen supplied in an air blast through the converter base, little or no scrap could be consumed, and the resulting steel had a high nitrogen content, a feature which was disadvantageous for certain types of deep drawing operations. In the open hearth processes, on the other hand, the principal role of gaseous oxygen was to sustain the combustion of externally supplied fuel, the oxidative refining processes being effected largely by the addition of iron oxide as ore or mill scale to the charge.

With the advent of pure oxygen in increasing amounts and at decreasing cost, it became evident that this might in principle be employed either to enhance refining, or to intensify combustion, or both. The simple substitution of pure oxygen for air in Bessemer converters should theoretically increase the rate of oxidation of impurities and consequently that of heat generation. In general, however, bottom blown Bessemer converters would not stand the thermal strain so imposed and, as a result, an oxygen enriched blast rather than pure oxygen was employed. Such enriched blasts reduce the ultimate nitrogen content and by increasing the rate of heat generation open up the possibility of utilizing some cooling scrap to achieve effective control and enhance the metal yield. Approaches to oxygen steelmaking of this kind were largely associated with basic Bessemer (i.e. Thomas) converters, and for this reason may conveniently be called oxygen-modified Thomas processes.

Derived from and related to this approach, there existed the possibility of blowing oxygen on to the molten metal from the top of a converter. This possibility was realized in a group of processes all involving top blowing of which the L.D., O.L.P. or L.D.–A.C., Kaldo and Rotor have emerged in distinctive forms. Taken together, these processes are the core of the revolution in steelmaking brought about by the availability of tonnage oxygen.

The prospect of using oxygen instead of air to intensify the heating in open hearth furnaces is a comparatively old idea, but the practice of using oxygen directly to speed up the refining operation itself was, in some measure, a reaction of open hearth steelmakers to the new converter technology. This development is considered under the heading of intensively oxygen-assisted open hearth processes.

A third group of processes are related to the use of oxygen in electric steelmaking with or without the addition of a fuel. Beyond these possibilities, there is the general prospect of carrying out some refining of hot metal in transit in ladle or mixer from the blast furnace to steelmaking furnace. If, for example, some degree of refining could be obtained by the simple expedient of blowing a jet of oxygen on to the hot molten metal in ladle or mixer without the need to establish elaborate facilities, such a practice could, in some situations, provide a means of decreasing the metallurgical load on the steelmaking furnaces and thereby enhancing the overall production rate. As this tends to be a marginal activity, it will be considered only briefly under the heading of prerefining with oxygen.

In summary, oxygen steelmaking processes may be classified as follows:

(i) Oxygen modified Thomas (basic Bessemer) processes.
(ii) Oxygen converter processes.
    (*a*) L.D.
    (*b*) O.L.P. or L.D.–A.C.
    (*c*) Kaldo or Stora-Kaldo.
    (*d*) Rotor.
(iii) Intensively oxygen assisted open hearth furnaces.
(iv) Oxygen assisted electric steelmaking processes.
(v) Oxygen prerefining procedures.

While those listed under (ii), (iii) and (iv) are of major importance, each group is now considered in turn.

OXYGEN-MODIFIED THOMAS PROCESSES

The intrinsic, long-standing disadvantage of Thomas steels in deep drawing operations arose primarily from their high nitrogen content which increased the tendency of the steel to strain age embrittlement. Though the possible use of oxygen instead of air had been envisaged by Bessemer, one of the early practical approaches towards partially overcoming this disadvantage of Thomas steels, introduced as early as 1925, involved the use of an oxygen enriched blast. After the Second World War, it became common practice in Thomas plants in Germany, France, Belgium and Luxembourg to use a blast containing 30 per cent oxygen which appeared to be the optimum concentration with the then existing converters. This produced two effects: (i) it permitted, on the average, 280 lb scrap to be included in the charge per 1,000 cu.ft. of oxygen employed; and (ii) it produced steel in which the nitrogen content was reduced from the normal value of 0·01 to 0·007 per cent.

If nitrogen were included in the blast, even in association with various special practices designed to minimize the residual nitrogen content, such as low blast pressures at high rates, a bath geometry providing minimum contact, minimum finishing temperatures and vacuum or flushing treatments, the nitrogen content of the finished steel still remained too high. This led to trials using nitrogen-free blasts composed of oxygen and steam and oxygen and carbon dioxide. With such blasts, a greater degree of freedom with respect to operating temperature was possible and, as a result, phosphorus contents could also be reduced below previously attainable levels. This was of some importance since this element also has deleterious effects on deep drawing properties.

The first published accounts of oxygen-steam trials carried out in Germany and Belgium appeared in 1950 and 1951, respectively. A fair variety of oxygen-steam compositions was examined, in addition to two stage blowing initially with oxygen enriched air followed by oxygen-steam. The subsequent choice of particular practice and of gas compositions employed was related to the composition of the molten metal, especially with respect to the phosphorus content, the bottom life of the converter, and the amount of scrap which was to be incorporated in the charge. Acceptable bottom lives with a 15 per cent scrap usage and a final nitrogen content averaging 0·001 per cent could consistently be achieved. The resulting steel was designated V.L.N. (very low nitrogen) and was in all respects similar to open hearth grades.

Studies on the use of oxygen-carbon dioxide mixtures date from the patents[1] taken out in 1943. The practice was vigorously followed up after the Second World War by Kalling and his colleagues in Sweden and also in Germany and Belgium. Again, both single and double stage blowing with oxygen enriched air followed by oxygen-carbon dioxide mixtures in several ratios around 1·2 : 1 were employed. The process proved to be flexible in operation and the finished steel had a nitrogen content in the range 0·004 to 0·006 per cent. It was, however, desirable to redesign the bottom of conventional converters to provide a somewhat smaller blowing area. This development was adopted on a production basis in Belgium, but, as with the oxygen-steam modification of conventional Thomas practice, it had little serious prospects of widespread usage or survival following the advent of top blown oxygen processes.

An important aspect of these developments is that they clearly demonstrate that, in the period before 1950, there had been extensive experimentation and considerable experience in these essentially evolutionary innovations derived from conventional practice. The results of this work were widely published and, if they served no other purpose, were at least a factor which preconditioned steelmaking opinion for the more revolutionary ideas that were to follow. In Sweden, the situation was more direct, in that Kalling, who had been involved in the oxygen-carbon dioxide work, was to be the principal architect of the Kaldo process described below.[2]

OXYGEN CONVERTER PROCESSES

The L.D. process. The L.D. process is usually thought to derive its name from the initial letters of two Austrian towns, Linz and Donawitz, where the first commercial exploitation took place. An alternative suggestion attributes the name to the initial letters of Linzer-Dusenverfahren (Linz jet). However, the process had its immediate origins outside Austria. The development of the Linde-

---

[1] T. Haglund, British Patent, 464,773 (1946); Swedish Patent, 124028 (1943).
[2] See p. 138.

Fränkl cycle during the period 1902–28, brought the extensive use of oxygen in steelmaking and in other spheres within the realm of near practical possibility. In the 1930's, Robert Durrer, Professor of Metallurgy in Berlin, widely and forcibly advocated the use of oxygen in metallurgical practice generally and supervised in Berlin from 1933 to 1939 experiments on the use of oxygen in various steelmaking techniques, including top blowing in a converter. The work was discontinued on the outbreak of war. In 1939, K. V. Schwarz, an employee of the Brassert Company in Switzerland, had patented[1] the use of a gas jet directed on to the surface of a bath of molten iron as a means of producing steel, but apparently was unable to produce material of acceptable quality. Another version of a similar idea had been patented in British Commonwealth countries by John Miles and partners.

After the Second World War, Durrer returned to his native Switzerland and became associated with the Gesellschaft der Ludwig von Roll'schen Eisenwerke A.G., at Gerlafingen, where, in 1947, he set up an experimental, three ton, top blown, oxygen converter. In March and April 1948, one of his associates, H. Hellbrügge, carried out a number of trials and showed that acceptable steel could be produced in ton quantities by this method and, furthermore, that some cold scrap could be added to the molten pig iron charge. These findings were not, however, taken up in Switzerland, but an account of the trials and the results were published in *Stahl und Eisen* in 1950 and, in this way, widely disseminated.

We now turn to the situation which was developing in Austria at this time. Austria had emerged from the war with a seriously unbalanced iron and steel industry, with iron production far outstripping steelmaking capacity. Supplies of coking coal which had to be imported from West Germany and Poland were in limited supply and likely to remain in this state. Scrap was not available in significant quantities owing to the comparatively small population and low level of industrialization in the country. In addition, the pig iron made in Austria from the Styrian ores contained $0 \cdot 2$ to $1 \cdot 0$ per cent silicon and $0 \cdot 10$ to $0 \cdot 15$ per cent phosphorus and had been shown as early as 1938 by Herzog working at Hamborn and later in 1943 by Trenkler at Hagendingen to be unsuitable for making ingot steel by bottom, air blown, Thomas practice. A further factor in the situation in Austria was a pressing need for a strip mill requiring steel of high

[1] K. V. Schwarz, D.R.P. 735,196; U.S. Patent, 33,388.

ductility and, by implication, low nitrogen content. In terms of conventional steelmaking, the building of an open hearth furnace seemed to be the only technical possibility, but many of the factors noted above were unfavourable to this solution and the capital cost of such an installation was high even for the nationalized Austrian industry.

Accordingly, in May 1949, a meeting of the Technisch-wissenschaftlicher Verein Eisenhütte Österreich, conscious of these considerations, resolved to embark on a co-operative development programme on the use of oxygen in steelmaking as an alternative to conventional processes in an attempt to resolve the Austrian problem. The works involved were the Vereinigte Österreichische Eisen und Stahlwerke Aktiengesellschaft (VÖEST), the Österreichische-Alpine Montangesellschaft (ÖAMG), the von Roll'schen Eisenwerke, Gerlafingen, and Hüttenwerke Huckingen AG. VÖEST was to be specially concerned with oxygen steelmaking from different sorts of pig iron, ÖAMG was to deal with problems relating to manganese contents, von Roll'schen Eisenwerke to concentrate on tests with oxygen in electric steelmaking and Hüttenwerke to examine the refining of basic Bessemer steels. The important first trials were begun at VÖEST in June 1949 by a team led by T. E. Suess, their works manager, and included H. Trenkler, who had joined the company after the war, H. Hauttmann, R. Rinesch and F. Klepp.

The starting point of these trials was evidently the work of Durrer and Hellbrügge carried out in Berlin and in association with von Roll'schen Eisenwerke and previously referred to. These workers had produced steel in one ton quantities by top blowing with oxygen in March 1948, and the VÖEST team could therefore approach their task with some measure of confidence. Their trials, extending over about six months in the latter half of 1949 using a 2-ton converter, were successful, and on 9 December 1949 a decision was taken to build an oxygen steelmaking plant at the VÖEST works at Linz. The particular contributions made by the team under Suess lay primarily in the positioning of the jet, the design of the nozzle geometry and the selection of the oxygen rates appropriate to the nature of the charge. Without detracting from the Austrian contribution, it is well to recognize that this development emerged from two interacting situations, the existing state of knowledge on oxygen steelmaking and the peculiar technical, commercial and national conditions that existed in Austria at that time.

The decision by VÖEST to build an oxygen converter plant at

Linz in December 1949, was followed shortly after by a similar decision by ÖAMG at Donawitz. The Linz plant became operational in November 1952, and that at Donawitz in May 1953. In both cases the vessel size chosen was 25–30 tons.

The essential features of the L.D. process were initially covered by Austrian patents and by the end of 1951 VÖEST held more than ten. ÖAMG patented separately on lance design, the addition of solid oxides, the use of specific surface pressures and on converter design. To facilitate the exploitation of these and later patents on the international scene, all of them were placed in the hands of Brassert Oxygen Technik A.G. (B.O.T.) of Zürich, Switzerland; this company has continued to hold the exclusive rights to the most important group of patents. By the end of 1956, B.O.T. held, on behalf of VÖEST, 24 Austrian and 132 foreign patents and, for ÖAMG, 22 Austrian and 108 foreign, as well as some 20 foreign patents arising, in part, from the earlier work of Schwarz and of Miles. Arrangements were also made between VÖEST and B.O.T. for the former to provide trial facilities for customers contemplating L.D. steelmaking.

Among the early companies to take up the process were the following:

Dominion Foundries & Steel Limited, Hamilton, Ontario, Canada.
McLouth Steel Limited, Detroit, U.S.A.
Jones and Laughlin, Pennsylvania, U.S.A.
Gussstahlwerk Witten A.G., Witten, Ruhr.
Gussstahlwerk Bochumer Verein A.G., Bochum.
Nippon Kohan A.G., Japan.

A list of L.D. steelmakers recently published[1] reveals that the process, either in its original or modified forms, is now operated in nineteen countries. The existing and immediate future situation is summarized in Table 13.

The L.D. vessel is not unlike a Bessemer converter in shape, but has a solid bottom. Operation is commenced by charging the vessel with a quantity of scrap, the hot metal and sufficient lime to give a final lime:silica ratio of about 3:1, together with any fluorspar and mill scale that may be required. With the vessel in an upright position, refining is carried out by directing a stream of pure oxygen (99·5 per cent) through a water cooled jet situated above the molten metal. In the original Austrian development with vessels of about 30 tons

[1] *Metal Bulletin*, no. 4933, 25 September 1964.

Table 13: World distribution of L.D. steel plants

Country	Existing annual capacity (Short tons)	Future additions to annual capacity (Short tons)
Algeria	—	990,000
Australia	1,280,000	560,000
Austria	1,980,000	—
Belgium	1,770,000	4,910,000
Brazil	1,050,000	680,000
Canada	3,100,000	—
Finland	—	550,000
France	3,370,000	—
Germany (Western)	7,710,000	2,310,000
Greece	300,000	Under construction
India	830,000	960,000
Italy	—	6,070,000
Japan	18,890,000	5,490,000
Luxembourg	440,000	580,000
Netherlands	2,500,000	—
Norway	500,000	—
Peru	—	Under construction
Portugal	350,000	—
Spain	440,000	550,000
Sweden	110,000	—
Tunisia	—	90,000
Turkey	—	520,000
United Kingdom	4,420,000	1,680,000
United States	16,640,000	19,900,000
U.S.S.R.	4,070,000	6,630,000
Yugoslavia	—	660,000
Total World-wide	69,750,000	53,130,000

capacity, the oxygen requirements were 2,050 to 2,130 cu.ft. per ton, but this figure has been raised in some subsequent operations. The blowing time averages 18 to 20 minutes, depending on the composition of the charge and the oxygen rate; the tap to tap time is of the order of 36 to 40 minutes. In some parts of the world, vessels as large as 200 or even 300 tons are now employed with about the same time schedule. The bare comparison in terms of rates of production between a 300 ton open hearth furnace operating on a time cycle of 10 to 12 hours with an L.D. vessel producing 200 tons of steel in under an hour requires little comment.

The L.D. process was originally developed to treat Austrian pig iron containing 0·1 to 0·15 per cent phosphorus. One of the reasons for its early exploitation in North America stemmed from the existence there of similar materials. In other places, the phosphorus content of the available pig iron was considerably higher, and it was not long before there arose a widespread desire to extend the process for such feedstocks. This development is the subject of the next paragraphs.

The L.D.–A.C. process. As early as 1956, Trenkler at Linz claimed to have established that the L.D. process, as such, could treat satisfactorily irons containing up to 0·5 per cent phosphorus by increasing the amounts of lime and/or limestone added to the charge. This was an obvious approach against the general background of steelmaking, but this simple modification would not permit irons of still higher phosphorus content to be refined. One difficulty in refining phosphoric irons is the need to produce a fluid lime slag sufficiently quickly and, if possible, without recourse to fluxing agents. During 1956–7, workers at Acières de Pompey in France began to explore the possibility of treating iron containing 2 per cent phosphorus in a Thomas converter modified for top blowing. Two particular points emerged: (i) the beneficial results from adding small lump lime, this being aimed at enhancing the rate of formation of a fluid lime slag; and (ii) the adoption of a double slagging procedure. The latter was not an uncommon practice in conventional steelmaking, the first slag, high in phosphorus, being removed and the second, too rich in iron to be discarded, being retained for the subsequent charge. A number of other variants of this theme tried at other places was directed, in particular, to the modification of the oxygen nozzle system to provide a double stream, one at high velocity to penetrate to the metal for refining purposes and the other at lower velocity primarily to promote slag fluxing.

The small lump lime used at Pompey pointed the way to the development separately in Luxembourg and in France of the idea of reducing still further the size of the lime and adding it to the charge more or less continuously in a finely divided form suspended in the oxygen stream. One set of trials along these lines which was conducted at the Acières Réunies de Burbach-Eich-Dudelange (A.R.B.E.D.) in Luxembourg in conjunction with the Belgian Centre National de Recherches Métallurgiques (C.N.R.M.) led to a process identified by the letters O.C.P. (Oxygène-Chaux-Pulvérisée). The other work was done simultaneously in France by the Institut de Recherches de la Sidérurgie (I.R.S.I.D.) and subsequently in works at Denain. This version of the process became known by the letters O.L.P. (Oxygène-Lance-Poudre).

In 1958, C.N.R.M. made an agreement with the L.D. patent holders, Brassert Oxygen Technik, and the O.C.P. process was renamed L.D.–A.C., A and C being taken, respectively, from the initial letters of A.R.B.E.D. and C.N.R.M. A similar agreement was reached with I.R.S.I.D., thus uniting these developments with the original Austrian work. There is, however, rather more to the L.D.–A.C. process than simply adding lime in the form of powder. In practice, a double slag procedure is employed, some 10 to 30 per cent of the total lime requirements are added at the outset of refining, and where a substantial amount of scrap is included, iron ore is also added to the charge. Special attention needs also to be given to the lance position and the injection of the lime powder. The preferred practice seems to be to commence with the probe in a high position and to dispense no lime during the first five minutes of the blow in order to form rapidly a foaming slag. At this stage, the probe is lowered and lime dispensing commenced. After a total blowing period of fourteen minutes, when the carbon has usually fallen to 0.7 to 1.0 per cent and phosphorus to 0.2 per cent, the slag containing about 20 per cent phosphorus pentoxide (P_2O_5) is removed. The second stage of blowing is then commenced following the addition of scrap or ore for temperature control, this final period of simultaneous addition of oxygen and lime occupying three to five minutes. The rates of lime addition in these two stages are of the order of 145–200 and 45–90 lb per ton, respectively; the particle size is usually less than 2 mm.

The L.D.–A.C. process has also been applied with advantage in refining irons containing only 0.2 per cent phosphorus, especially where the silicon content is unduly high. In such cases, no intermediate deslagging is necessary, but the use of a foaming rather than a non-foaming slag has significant advantages in respect of phosphorus removal, final nitrogen content, lime consumption and metallic yield. It seems likely that in the future most plants of the L.D.-type will be provided with facilities for dispensing small lump or powdered lime, whether the immediate requirements are to treat high or low phosphorus irons, and best practice may advantageously take greater account of the desirability of controlling the dispersed nature of the slag.

There are a number of general points which apply more or less equally to the L.D. and L.D.–A.C. processes. The first is that top blown steel tends to be rather higher in manganese than the comparable open hearth product, but this is seldom disadvantageous for many applications. The L.D. and L.D.–A.C. processes are also effective in reducing the sulphur content, but, as the final sulphur level is much dependent on the concentration of this element initially present in the charge and, in any event, preliminary desulphurizing with soda ash is well established in conventional practice, this is not a very special advantage of top blown steel unless a very low final sulphur content is required. It is also generally believed that tramp elements, such as arsenic, lead and zinc, are kept at lower levels in L.D. and L.D.–A.C. steels because they tend to evaporate at the rather higher temperatures employed compared with those in conventional steelmaking processes.

Without considering the question in detail at this point, it suffices to say that the capital costs of L.D. process equipment is in total less than that for comparable open hearth facilities. However, there is the major additional requirement of providing for fume elimination. One characteristic of L.D.-type processes is the dense red fume, largely of iron oxide, which is generated during oxygen blowing. This represents a loss of metal and, if not treated, would create a nuisance which clean air acts, in general, would scarcely tolerate. The carrier gas is very hot and contains a large amount of carbon monoxide which has a significant value as a fuel. Cooling of the gas as a necessary preliminary to cleaning is usually carried out by means of a waste heat boiler, but such a unit built to utilize an intermittent source of heat from a batch process will also have to be supplied with another method of heating in the off-peak periods. In consequence, it is likely to be awkwardly expensive in capital. Precipitation facilities also involve substantial capital expenditures, and these are charges which L.D.-type plants have to bear. An additional requirement is for pelletizing equipment to permit reuse of the precipitated oxide, though this may in some cases be more conveniently recycled to an existing sinter plant preparing blast furnace burden.

A further matter of importance is the life of the refractory lining of the converter. This is influenced by both mechanical factors, especially where substantial scrap additions are made, and by the composition and character of the slag, reflecting both the composition of the charge and the particular slag management that is adopted. Magnesite, dolomite and tar-dolomite linings have been used. Refractory lives vary considerably for the reasons noted above; in Japan and Austria, as many as 400 heats are said to be achieved, whereas in other places 200 are often considered normal. Since 200 heats represents about a week's operation, it is obvious that for every L.D.-type installation, at least two vessels are required, one in use and one undergoing relining. The more intensive conditions existing in oxygen steelmaking vessels in comparison with conventional furnaces and converters has pointed up some lag in refractory technology and, incidentally, has given Austria a special advantage in the high quality of its naturally occurring magnesite deposits. There are at least two approaches to this problem: in one, the aim is to extend lining life and reduce relining time by using large preformed shapes of high quality material; in the other, the aim is to treat the lining as a short term consumable, and, therefore, to seek a cheap, rapidly and largely mechanically applied lining.

Finally, there is the question of scrap utilization. Originally, the L.D. process was based on a situation in which scrap was in short supply, but, as the technology developed and the range of pig irons refined by this process increased, scrap and iron ore were added to the charge in increasing quantities as a coolant. In so far as iron ore is used, the process becomes one of direct reduction, the metallic yield being thereby enhanced. In this way, both the L.D. and L.D.–A.C. processes began to acquire greater flexibility with respect to the feed than had originally been conceived and began to move some way towards conventional open hearth practice. Certainly, the capacity of these oxygen processes to utilize scrap, even to the extent of 25 per cent, is an additional advantage over air blown Bessemer converters.

The Kaldo process. The Kaldo or Stora-Kaldo process derives its name from that of Professor Bo Kalling and the works in Sweden, Domnarvet Jernverk, which is a branch of Stora Kopparbergs Bergslags A.G., where its development took place. Its earlier origins in the 1940's stemmed from a recognition that steelmaking, in general, might be improved in rate if the charge could be stirred to enhance the rates of reaction and/or mass transfer between the several phases. The principal work involving the use of oxygen began in 1948 and the decision to install a production unit at Domnarvet was taken in 1954. Noticeably, this period also spans that during which the L.D. developments in Austria were being most actively pursued. The particular feedstock for which the Kaldo process was devised was not the low phosphorus iron of Austria, but a material containing $1 \cdot 8 – 2 \cdot 0$ per cent phosphorus which was already being handled by Thomas converters in a quite standard way. In this respect, the Kaldo process is more a competitor of L.D.–A.C. than of L.D., but in many other respects it is often more appropriate to look upon it as an articulated open hearth furnace capable of high and flexible production rates.

The operating vessel—initially of 30 tons capacity—in the form of a cylinder coned at one end is mounted so that it can rotate about its axis at up to 30 revolutions per minute and be inclined to the horizontal. The oxygen is supplied through a water cooled lance inserted through the exhaust hood and is blown against the surface of the bath at a shallow angle which can be varied considerably. During operation, the furnace is inclined at 17° to the horizontal with the charge covering at least half the back wall. With Thomas irons, a two slag practice is followed, the first slag being removed when the phosphorus has fallen to 0·2–0·3 per cent. Lime, iron ore and scrap additions are made as required to promote the refining reactions and to achieve a smooth, controlled operation.

A feature of the Kaldo process is the number of variables which, in principle, may be controlled by the operator. For example, in the initial boil during removal of silicon, the action may be controlled by varying the rotational speed, or by altering the angle of impingement of the oxygen. During dephosphorizing, the state of the slag can often be gauged from the load on the driving motor which decreases when the slag liquefies. There is also the possibility, to an extent greater than in the L.D.–A.C. processes, of being able to remove unwanted impurities more selectively and under greater control, thereby decreasing to some degree the need for recarburizing. This is an important consideration when 'catch carbon' techniques are being practised.

A further feature of the Kaldo process is its ability to burn within the vessel most of the carbon monoxide evolved during the refining of the charge. The heat so generated may be transferred to the charge through the vessel lining as a consequence of rotation of the vessel, with the result that quite large quantities of scrap (as high as 35 per cent) may be added. The gases finally evolved are low in carbon monoxide and there is a marked decrease in the capital needs required to treat this effluent in comparison with the L.D. and L.D.–A.C. processes.

The time cycle of 1·5 to 2 hours from tap to tap with a direct refining period 45 to 50 minutes is longer than that with two slag L.D.–A.C. practice, but this is offset to some extent by rather greater flexibility and potentially enhanced control. An outstanding disadvantage is, however, the much shorter refractory life. Although figures as high as 100 to 200 heats have been obtained, the average appears to be in the range, 60 to 70, with some low figures in the 40's, compared with figures in excess of 200 fairly regularly obtained with L.D.–A.C. vessels. Wholly valid comparisons of this kind between processes are not easy to obtain, since the nature of the irons being treated, the quality of the refractory and the detailed slag practice vary from case to case. It is, however, generally agreed that Kaldo vessels have shorter refractory lives under otherwise comparable conditions arising from the moving charge and the extensive utilization of the lining for heat transfer purposes. Oxygen usage which is about 2,000 cu.ft. per ton of steel is a little higher than the figure of 1,700 to 2,000 cu.ft. usually quoted for L.D. and L.D.–A.C.

Some engineering problems are posed by the rotation of the vessel, especially where particularly large units are installed. The bearings have to carry a very high load and operate under severe conditions of elevated temperatures. Moreover, the shell of the vessel itself and its appurtenances have to perform functions other than simply enclosing the refractory. Distortion of the vessel, for example, can pose serious difficulties. Undoubtedly these problems and deficiencies can be overcome, but not without expenditure of time, effort and money.

Nominally, the Kaldo process is a competitor of the L.D.–A.C. and open hearth furnaces, but there is also the growing prospect of using the L.D. process with substantial lime addition and two slag practice to treat high phosphorus irons. The Kaldo process is capable of producing under good control steel to almost any final specification from iron of almost any phosphorus content with a comparatively high usage of scrap and iron ore. Whether the higher capital cost, engineering problems and higher refractory usage will ultimately enable it to compete effectively with L.D., L.D.–A.C. or modifications of these processes appears rather doubtful. In spite of this, the process has found considerable favour in its native Sweden, in the United States, France and the United Kingdom. Details are shown in Table 14.

The Rotor process. The development of this process took place initially at Hüttenwerk Oberhausen in the Ruhr and is variously known as the Rotor, Graef Rotor or Oberhausen-Rotor. It arose against a background in which there were two main features: (i) the general situation in the world steel industry with respect to oxygen steelmaking in the late 1940's and early 1950's; and (ii) the particular conditions which existed in the Ruhr. Unlike the situation in Britain

Table 14: Kaldo plants in operation and under construction on 1 October 1965[1]

Operating from	Steelworks	Furnace Capacity Metric tons	Hot Metal	Coolant	Main Products
1956	Domnarvet, Sweden	30 × 1†	High P (1·8)	Ore	Plate, sheet, wire.
1960	Sollac, France	130 × 1†	High P (1·8)	Scrap	Deep drawing sheet.
1961	Oxelösund, Sweden	130 × 2	Low P (0·1)	Ore/ Scrap	Plate.
1962	Sharon, U.S.A.	140 × 2	Low P (0·1)	Scrap	High and low carbon. Alloy steel.
1964	Park Gate, England	80 × 2	High P (1·3)	Scrap/ Ore+ Cold Pig	High and low carbon. Alloy steel.
1964	Consett, England	120 × 2	Med P (0·2)	Scrap/ Ore	High carbon.
1964	Shelton, England	50 × 2†	High P (1·3)	Scrap	High and low carbon (continuous casting).
1964	Norrbotten, Sweden	70 × 1	High P (1·8)	Ore	Sheet and medium carbon.
1965	Stanton, England	70 × 1†	High P (1·8)	Ore/ Scrap	High grade foundry iron.
1965	Domnarvet, Sweden	80 × 2†	High P (1·8)	Ore	Plate, sheet and wire.

or the U.S.A., a large proportion (over 40 per cent) of the steel made in Germany was refined in Thomas converters. In these years there was an increasing demand for steels of open hearth quality to meet a changing pattern of requirements. One possible approach to the achievement of higher throughputs in existing open hearth furnaces in the short term was to pre-refine pig iron charges intended for open hearth operations. This has the two advantages of minimising the variability of the ultimate feed and reducing the metallurgical load on the steelmaking furnaces.

The Rotor furnace was, in these circumstances, initially conceived as a pre-refining unit, but almost from the beginning it was realized that it was inherently suitable for steelmaking. The particular advantages claimed were its superior desulphurizing characteristics and its capacity to produce steels low in phosphorus and nitrogen equivalent to the best open hearth qualities. The initial trials at Oberhausen were made in a 60 ton vessel and the first commercial unit of 100 tons capacity was commissioned in 1958.

The Rotor furnace is essentially a long cylindrical furnace mounted

[1]Data supplied by Stora Kopparbergs Bergslags Aktiebolag.
† With exchange vessel.

axially and provided not with one but with two oxygen lances. The primary lance, which is submerged in the charge, serves to introduce oxygen to oxidize impurities and to promote turbulent slag-metal mixing, thereby enhancing the rate of mass transfer between these phases. The second lance, which is fed with less pure oxygen or oxygen enriched air, leads above the charge and serves to promote combustion of the carbon monoxide within the furnace, the heat so generated being transmitted to the charge via the furnace lining, and to contribute to the refining operation by contact with the turbulent slag-metal charge. The vessel may be rotated about its axis at low speeds, $0.1-0.5$ r.p.m.—which is much slower than the Kaldo—and powdered additions of fine ore, lime and limestone may be made continuously during the blow. For mechanical rather than metallurgical reasons, it is comparatively inconvenient to utilize much scrap, unless this is in specially selected sizes; customarily, up to 10 per cent is used.

This process clearly embodies a high degree of operating flexibility. The two oxygen lances with programmed flows, the ability to vary the position of the lances, especially the depth of immersion of the primary lance, and the ability to feed solid additions more or less continuously are features of the process. The rotational speed, small in absolute terms, is not particularly significant as a variable influencing the rate of the refining reactions, but contributes to the improvement in the heat transfer between the lining and the charge.

The symmetry of the vessel, its slower rotational speed compared with the Kaldo and the possibility of varying the disposition of the lances during a campaign are features which would be expected to give superior refractory lives. The number of heats possible with a given lining varies according to the particular practice followed and, because of the relatively small number of furnaces in commercial operation, average figures are not very meaningful. Some eighty heats are said to be usually possible before the working lining has to be replaced. The cycle time on a Rotor furnace is often quoted with an average of 2 hours 10 minutes, though shorter times have been used in some plants. The Rotor furnace necessarily poses similar engineering problems to the Kaldo in respect of its rotational requirements and had the additional restriction that, if scrap is to be included in the charge, this must be specially selected and prepared in order that it can be added through the rather small opening.

Early installations of Rotor plants have included 100 ton units at the Vanderbijl Park Works of I.S.C.O.R. in South Africa (1959), the Peine Works of Iseder Hütte in Germany (1959), and at the Redbourn Works of Richard Thomas and Baldwins at Scunthorpe, U.K. (1961).

INTENSIVELY OXYGEN ASSISTED OPEN HEARTH PROCESSES

When pneumatic oxygen steelmaking processes described in the previous sections began to make their appearance, initially to meet specific local conditions, but soon to exhibit potential on a wider scale, more than three-quarters of the world's steel was made by open hearth furnaces. It is scarcely surprising, considering the capital investment in these furnaces, that a large number of open hearth works in many parts of the world began about 1950 to seek ways of making use of tonnage oxygen in open hearth practice. There are two potentially useful areas: (i) the enrichment of combustion air; and (ii) the direct acceleration of refining reactions.

The use of fuels of higher calorific values coupled with the use of oxygen enriched air permits a higher rate of heat input to the open hearth furnace thereby reducing the melt-down time, which is a significant component in the total cycle. Moreover, oxygen added in this way can increase the oxygen potential of the slag resulting in somewhat enhanced refining rates. Primarily, this approach involved issues of nozzle and checker design, but did not alter fundamentally the steelmaking process or practice.

The second area of more direct use of oxygen to accelerate refining was also taken up enthusiastically. Initially, this involved the lancing of the bath with oxygen supplied through consumable lances inserted through furnace doors and ports without a serious attempt at proper engineering design, or full regard to the task of the operator. In Japan, in particular, this approach has enjoyed a long and fruitful popularity, possibly because of the enormously rapid expansion of the industry and less vigorous trade union practices. The next stage was to consider how probes could be built into the furnace through back walls or corners, but this was soon succeeded by attempts in 1953 to introduce retractable water cooled jet probes through the furnace roof. By 1959, this practice had been widely adopted, the general procedure being to lower the jet to within a few inches of the surface of the molten charge and to blow through a tip bent at an angle or through a multiple hole system. Widely varying oxygen flow rates were employed, depending on the composition of the charge and the rate of refining required, with figures as high as 1,000–1,200 cu.ft. per minute commonly being achieved. In the beginning, this development presented few apparent difficulties; the rate of sulphur removal was increased by the enhanced agitation and a substantial decrease in tap to tap time was obtained with some savings in fuel consumption, but without deleterious effects on steel quality.

While the early oxygen rates employed in roof jetting were around 125 to 150 cu.ft. per short ton per hour, this was soon increased to levels twice as great. Further developments embodied the use of multijet probes and the provision of facilities for introducing powdered solid ingredients, iron oxide, lime, limestone and fluxes along with the oxygen stream. Roof jetting practice adopted a variety of detailed forms in different parts of the world and was frequently coupled with the use of oxygen enrichment in the combustion air. There was to this time minimum modification of the furnaces and, though these various practices produced to a greater or less degree the desired improvements in production rates, these gains were frequently obtained at the cost of higher rates of refractory wear. In particular, the traditional silica roofs were unable to stand the enhanced thermal strain, a feature which led to the introduction of a wholly basic roof, but, perhaps of more importance, attention began to be directed towards the possibility of redesigning the open hearth furnaces positively to meet the demands of the intensive use of oxygen.

One of the most successful of these developments was the Ajax hot metal process introduced in 1957–8 by the Appleby-Frodingham Steel Co., a branch of The United Steel Companies Limited, at Scunthorpe in Lincolnshire. The name, Ajax, now attached to the furnace, is derived from that of Mr A. Jackson of that company who was responsible for its development. In its essentials, a large tilting open hearth furnace was specifically modified to permit oxygen to be efficiently utilized in three ways: (i) through the burners with coke oven gas; (ii) through water cooled probes; and (iii) through regenerators with combustion air. A new method of basic roof construction was also introduced to meet the intensified conditions in the furnace.

A brief description of the method of operation is as follows: Following a tap after which some 20 to 30 tons of molten metal and all the finishing slag from the previous heat are retained, the furnace is fettled with dolomite and charged with ore, and lime or limestone. During this period the furnace is fired with coke oven gas and oxygen is admitted through the burners to accelerate combustion. The molten iron is then charged from the mixer and, as soon as sufficient metal (only part of the total charge) has been added, the fuel is cut off and oxygen blowing begun through the probes at rates up to 60,000 to 75,000 cu.ft. per hour. At the same time, hot air is admitted through the regenerators to the extent of about 5·5 times the volume of oxygen being used. On completion of charging and following analysis of the hot metal, oxygen, lime and ore additions are made to give a bath containing 1 per cent carbon and 0·1 per cent phosphorus before the first slag is run off by tilting the furnace body. Further lime additions are then made and the necessary oxygen (at rates up to 90,000 cu.ft. per hour) and, if required by temperature considerations, some ore as well are added to bring the bath to about 0·3 per cent carbon above the final specification of the steel being made. Once the required amount of oxygen has been added, oxygen injection is stopped and fuel firing recommenced until tapping.

This brief description illustrates how far a modified open hearth operation of this kind has incorporated many of the features of oxygen converter practice whilst retaining the size, flexibility and control traditionally achieved in open hearth technology. The particular practice described is one based on hot metal, this being the process at Appleby-Frodingham, but it is, in principle, capable of wide variations with respect to feed. Although almost all the existing tilting open hearth furnaces at Appleby-Frodingham have now been successfully adapted for Ajax operation, and the process is available for purchase on comparatively modest terms, there have to date been no other recorded cases of its adoption. In Britain at least, this reflects, among other things, the fact that few other works have tilting open hearth furnaces, so that the potential gains obtainable from the low capital cost of conversion of existing tilting furnaces (£150,000–£200,000 per unit) cannot so readily be achieved.

OXYGEN ASSISTED ELECTRIC AND FUEL PROCESSES

While electric steelmaking furnaces have a special role to play in the production of alloy steels, it is difficult in isolation to conceive any situation in which they could become the dominant, large scale method of producing general purpose mild steel. Nevertheless, their particular advantage of being able to operate on a total scrap charge gives them a specific place in the total pattern of steelmaking processes, especially where electricity can be obtained at an acceptable cost. The Brymbo Steel Works of the G.K.N. Steel Co. Limited and the Steel, Peech and Tozer Works of The United Steel Companies Limited are two examples in Britain in which, within their respective groups of associated companies, works have deliberately been converted to electric steelmaking.

The use of oxygen in refining in electric steelmaking involves no new concept, except that the extent of the refining required with an all scrap charge is often considerably less than for a conventional pig iron in a normal thermal process. Oxygen jetting or lancing have obvious applicability. One problem here arises in fume removal since it will seldom be worthwhile to provide for this on anything like the scale for L.D., L.D.–A.C., Kaldo or Rotor processes. A second problem may be posed by enhanced roof refractory wear.

This situation, and the fact that electric power can seldom be obtained at an acceptable price except in large scale, multi-unit electric works, has led the British Iron and Steel Research Association to consider the possibility of using an oxygen-oil burner to achieve rapid melting and refining. This is the basis of the B.I.S.R.A. Fuel-Oxygen-Scrap process (F.O.S.). Of special interest here is the ancillary observation by B.I.S.R.A. that there is virtually no fume during lancing if the stream consists of oxygen that has been partially combusted with oil, so that it contains about 30 per cent excess oxygen.

These developments embrace a series of ideas and practices which could be combined in several ways in electric, electric-cum-F.O.S. and pure F.O.S. steelmaking to take the maximum advantage of the availability of scrap, the needs of particular markets both in respect of scale and steel quality, and the relative costs of oil fuel and electric power in relation to the utilization of oxygen in refining operations.

OXYGEN PRE-REFINING PROCESSES

In an ideal world, pig iron coming forward for steelmaking would have a uniform composition and be available in quantities which always exactly matched the steelmaking demand. There are, however, few if any situations in which these conditions prevail and there are distinct advantages in inserting large (1,000 ton) mixers between the iron making and steelmaking furnaces. Such mixers may be of two limiting types: (i) inactive, in which no positive effort is made to carry out any pre-refining; and (ii) active, in which some pre-refining is attempted primarily as a means of decreasing the metallurgical load and thereby potentially increasing the throughput of the principal steelmaking furnaces. It is also possible to carry out some pre-refining in transfer ladles where no mixer is employed and, indeed, the case for pre-refining is often stronger in such circumstances to reduce some of the variability in pig iron quality.

The practice of reducing the sulphur content of pig iron by the addition of soda ash to the mixer or ladle is a well established one. Pre-refining with respect to silicon and phosphorus are amenable to oxygen treatment through a lance with or without the addition of slag forming materials, such as lime or limestone. Control can be tolerably crude and it is customary to provide a minimum of ancillary facilities for fume removal. With the exception of soda ash treatment, it is probably better in the long term to improve blast furnace and steelmaking practice and to balance the relative capacities rather than to interpose additional pre-refinement procedures. Nevertheless, they exist, they require comparatively little capital and they can be effective, especially in the short term, in improving overall production rates. There is one situation in which pre-refining may be more readily justifiable. This is where the pig iron happens to be abnormally or irregularly high in such elements as chromium which could seriously upset the refining operation.

In recent years, a number of attempts have been made to bring pre-refining processes under more detailed control, or to devise more effective methods. Often, these have taken the form of spray refining by dropping molten pig iron in droplet form down a shaft furnace supplied with a countercurrent stream of oxygen. This approach offers some prospects of development into a near-continuous steelmaking process, and efforts are currently being directed in the United Kingdom to this end. In addition, it should be recalled that both the Kaldo and Rotor processes were conceived as or developed from pre-refining processes and that Bessemer converters offer a possible means of pre-refining to an extent which at various periods in the past has given rise to regular duplex practice.

The case for expenditure of capital on pre-refining facilities depends on a quantitative appreciation of a wide range of factors which apply in varying degrees to individual situations. It is, therefore, difficult to make any valid overall pronouncement upon it. In general, however, the usage of oxygen in this area has not increased substantially in recent years in line with the total pattern of oxygen usage and it would appear that oxygen pre-refining will be restricted to particular situations rather than develop towards universal application.

An assessment of the future of oxygen steelmaking

Steelmaking processes do not exist in isolation, but are bounded on one side by ironmaking and on the other by the wide variety of shaping, rolling and forming processes which yield the semi finished products of the industry. In this setting, any assessment of the future of oxygen steelmaking often has to be in particular terms rather than general ones. Nevertheless, some broad features emerge. The first is that oxygen steelmaking is clearly now well established and considerable expansion can be confidently predicted during the next decade. For example, in Britain the Iron and Steel Board[1] has estimated that the proportion of steel made in open hearth furnaces will fall from 76 per cent in 1963 to about 50 per cent in 1970 and that by that date pneumatic processes will account for 34 per cent of the production. Furthermore, L.D.-type processes currently account for the production of about 70 million tons of the world's steel and there are firm plans for future additions of 53 million tons to this figure.

The future of intensively oxygen assisted open hearth furnaces contains some uncertainties. For a time at least and especially where there exist large modern furnaces capable of adaptation, it will often prove more economical to adopt intensive oxygen practice to obtain additional production up to and possibly somewhat beyond 25 per cent of currently obtained conventional outputs. Whether the approach will be to remodel on the Ajax pattern or whether to evolve in steps the most suitable arrangement for a particular installation depends greatly on local factors. On the whole, the second alternative seems to be the more favoured one.

There is a fairly wide body of opinion which holds that completely new open hearth furnaces are unlikely to be built in preference to one of the converter units. The exceptions to this general rule are likely

[1] *Development of the Iron and Steel Industry*, Iron and Steel Board Special Report, 1964, H.M.S.O., p. 119.

to be few and possibly outstanding; for example, the development of the very large open hearth furnace by the Steel Company of Canada has now almost reached the stage of being more properly described as a semi-continuous converter rather than an open hearth furnace. Other developments of the same general kind include the twin hearth furnace.

Of somewhat greater interest is the possible competition between the four converter processes. Until fairly recently, the choice between L.D. on the one hand and L.D.–A.C., Kaldo and Rotor on the other rested primarily on the phosphorus content of the pig iron being used. This distinction is disappearing as techniques develop to enable two slag practice with lime additions to be adopted with L.D. plants. In some places, L.D.–A.C. plants originally operating as designed with powdered lime addition through the lance have turned over to more conventional L.D. operation with triple nozzle oxygen lances and separate lump lime addition. The distinction between L.D. and L.D.–A.C. appears, therefore, to be less marked than hitherto, and in future it will be more realistic to combine these practices within the wider framework of L.D-type.

There is no doubt that Kaldo and Rotor installations are more expensive in capital, pose additional engineering problems and have lower outputs than an L.D.-type plant of comparable size. Moreover, there are fewer problems in increasing the size of L.D. vessels to, say, 500 tons than would be the case with either the Kaldo or Rotor. Of the two rotational processes, the present pattern heavily favours the Kaldo and it appears as if the Rotor will not find any wider favour in the future. The choice reduces therefore to L.D.-type or Kaldo. The slower working of the Kaldo has to date offered, at least in principle, better prospects of achieving the degree of control necessary for a variety of steelmaking practices often related to the growing need in the industry to make an increasing number of rather closely specified steels. Extensive experience with and without computer assistance or control should ultimately permit almost any practice that might be required and any specification within the normal range of common steels to be met with L.D.-type plants working at normal rates. If this proves to be the case, the future of converter processes appears to be heavily in the direction of L.D.-type in preference to rotational units.

A further consideration is the consumption of scrap. Many observers take the safe line of predicting that the future lies in mixed processes—oxygen converters coupled with electric or some form of fuel-oxygen-scrap steelmaking. In this connection, it is worth noting that the proportion of a steel ingot which is recycled as works scrap has remained extraordinarily high and there are important economies to be made in this area. Continuous casting goes some distance in alleviating this problem, but there are likely to be continuing supplies of external scrap in most industrialized countries. In the longer term, if the world's ore reserves become progressively more difficult and expensive to exploit, scrap seems likely to have an increasingly favourable cost margin over hot pig iron. Indeed, the growth in the use of electric steelmaking for general purpose steels as distinct from special alloys is some indication of future trends.

The overall picture with respect to the principal materials required in the various steelmaking processes, both conventional and oxygen, are summarized in Table 15.[1] The figures quoted are at best typical, the actual quantities used varying appreciably on either side according to local conditions. Unfortunately, it is not possible to set down in such concise form on a wholly comparable basis the capital costs of these various types of plants. There have, of course, been attempts to arrive at comparisons both in respect to capital and operating costs, but these have often been related rather specifically to a situation preconditioned by existing facilities, or where they refer to the green fields case, are seldom sufficiently detached or broad enough in compass to be advanced with confidence. It would appear that each situation has to be determined separately.

Finally, we mention two matters which could modify quite drastically a substantial area of the iron and steel industry embodying the central steelmaking function. The first of these is the progressive development of automatic control and the second, the prospect of continuous steelmaking. Broadly speaking, the present situation is that rapid methods of chemical analysis for nearly all the elements significant in steelmaking are well established and efforts are being directed towards a better appreciation and assessment of the chemical composition and physical state of the slag. Most major steel companies are exploring with computers data collection and mathematical modelling as a prerequisite to setting up operational and control programmes, but the fully automated process regularly operated under computer control still appears to be some way off. There is a number of major scientific and technical problems in this

[1] *Development of the Iron and Steel Industry*, op. cit., p. 163.

Table 15: Examples of the principal materials required to make 1 ton of ingots by various steelmaking processes

Material	Traditional basic Bessemer steel plant	Open hearth processes				Electric arc processes		Basic oxygen converter processes			
		Cold metal furnaces	Hot metal fixed furnaces	Hot metal tilting furnaces	Ajax hot metal tilting furnaces	Special quality alloy steels	Common quality steels	L.D.	L.D.-A.C.	Kaldo	Rotor
Molten iron (cwt)	21–23	—	11	15·5	21	—	A little used occasionally	17–20·5	17–21	14–20	20–22
Cold pig iron (cwt)	—	6	A little used occasionally	—	—	0–1·5		—	—	—	—
Scrap (cwt)	0–1·6	16	10·5	5·5	0·2	20–22	21·5	2–6	Up to 6	Up to 8	Up to 2
Ratio of scrap to iron	Low	73/27	52½/55	25/77½	Very low	Very high	Very high	10/103 to 30/85	10/103 to 30/85	10/100 to 40/70	Low
Oxidising ore or millscale (lb)	Generally low but up to 130 lbs at times	45	200	300	175	5–35	25	15–150	15–220	75–330	25–330
Fluxes (lime and/or limestone and dolomite) (lb)	250–400	300	400	300	185–200	50–150	110	120–160	220–270	220–300	150–330
Finishing (ferro alloys) (lb)	22	28	28	20	20	Varies according to alloys	20	12–13	14	14	14
Oxygen (cu.ft.)	—	(a)	(a)	(a)	1,380	(a)	150	1,700–2,000	1,800–2,000	2,000	1,500–1,900
Primary fuel: Liquid gas (therms)	—	60–65	40–50	35–50	11	—	—	—	—	—	—
Electricity (kwh)	—	—	—	—	—	650–750	500–550	—	—	—	—

(a) Appreciable quantities but very variable.

area that have yet to be adequately tackled and solved, though progress is rapid, especially in Japan.

The second area of continuous steelmaking is still largely in the speculative stage, though a number of experiments in launder and other furnaces using oxygen has been conducted. If blast furnaces are made large enough they can be tapped more or less continuously, and either by substantial further development of the idea of continuously refining iron to steel by oxygen blowing as the iron passes under gravity from the blast furnace to a continuous casting plant, or, as seems more likely, by the evolutionary development of L.D.-type plant to give quasi-continuous operation, there is little doubt that, given sufficient effort, continuous steelmaking is not a wholly inconceivable achievement. It is, however, problematic whether there will be sufficient technical or commercial incentive to move strongly in this direction. There are, among others, two particular arguments against it: one is the increasing number of more specific steels that is being called for, a fact which militates against the very long production runs much desired in fully continuous processes; the other is that, as with most nominally continuous chemical processes, ballast or holding vessels or facilities to purge process streams to waste are normally required. This presents comparative little difficulty or cost in handling gases, liquids and even solids under the moderate conditions common in chemical processes. The situation is, however, a little different in dealing with hot metal and slag streams. Undoubtedly, there will be a sustained move towards quasi-continuous processes, but fully continuous processes in steelmaking may not in practice prove to be worthwhile. The use of L.D.-type converters with cycle times a good deal less than ten per cent of conventional open hearth furnaces and of not much more than this figure in comparison with intensively oxygen assisted open hearth furnaces is an important step towards quasi-continuous steelmaking.

The economic, political and social background of the British steel industry

It is scarcely necessary to emphasize that in any heavily industrialized country steel occupies a central economic position. This is specially true of Britain which for generations has supplied many parts of the world with manufactured goods and machinery made primarily from steel. Moreover, anyone who has lived in Britain at any time in the past thirty years cannot be unaware of the internal social and political aspects of the iron and steel industry. For quite obvious reasons, these aspects cannot be ignored in any study of a major technical innovation in steelmaking. There is little purpose to be served in considering the situation before the First World War, and it is from this point that we shall take up the narrative.

During the First World War, governmental requirements for increased production of shell and armament steels had inflated productive capacity, all of which was neither required nor well adapted to postwar needs. In 1918, production of steel in the United Kingdom was 9·54 million tons and, in the period immediately following, short lived optimism resulted in the installation of further steelmaking plant. After a fall to 7·89 million tons in 1919, production again rose to 9·07 million tons in 1920, but from then on the twenties and thirties were a period of depressed uncertainty for the British steel industry. The total picture is illustrated by the figures given in Table 16.

Table 16: U.K. steel supplies and consumption, 1918–31
(in million tons)

Year	Production	Imports	Exports	Apparent home consumption
1918	9·54	0·24	1·27	8·50
1919	7·89	0·37	2·15	6·12
1920	9·07	0·99	3·25	6·81
1921	3·70	1·46	1·69	3·47
1922	5·88	0·79	3·39	3·28
1923	8·48	1·18	4·31	5·35
1924	8·20	2·17	3·84	6·52
1925	7·38	2·59	3·36	6·61
1926	3·60	3·55	2·69	4·46
1927	9·10	4·19	3·78	9·51
1928	8·52	2·75	3·83	7·44
1929	9·64	2·68	3·94	8·38
1930	7·33	2·82	2·92	7·23
1931	5·20	2·76	1·83	6·13

The slump in production in 1921 was accompanied by a sharp fall in price and, though there was a marked recovery in the years 1922 to 1925, production again fell to a record low level of 3·6 million tons in 1926. This was followed by a further short-lived recovery preceding a further slump in the early 1930's, so that it was not until 1935 that crude steel production exceeded the level of 1920. The import picture is also revealing, particularly the substantial growth till 1927 and the stabilization thereafter at around 2·8 million tons. The wildly fluctuating character of exports is an index of the complex problems that beset the industry in these years.

The steel industry is a heavily capitalized one and standing charges accumulate whatever the level of production. There is, therefore, a universal desire to maintain levels of production as high as possible in order to spread these charges. But the demand for steel cannot in the short term be increased simply by a reduction in prices. In a period in which productive capacity continues grossly to exceed demand, price cutting can go on only as long as the prices continue to exceed direct operating costs, even though they make an inadequate contribution towards meeting the standing charges. If such a pattern develops strongly, it is inevitable that the industry ultimately fails to pay its way, it is unable to attract fresh capital and it progressively becomes more and more technically out-of-date. This was roughly the situation of the British steel industry in the 1920's and early 1930's.

One of the first effects was to promote reorganization and capital reconstruction of some of the largest firms, in some cases under the influence of the Bank of England of which many of the major companies were customers. In 1927 Vickers merged with Armstrong-Whitworth to create the English Steel Corporation; in 1930, the Lancashire firms of Pearson and Knowles, Rylands Brothers and Parkington Steel and Iron Company combined to form the Lancashire Steel Corporation; Baldwin's heavy steel interests at Port Talbot were merged with Guest, Keen and Nettlefolds at Dowlais and Cardiff to form Guest Keen Baldwins. In 1930, The United Steel Companies were reconstructed under a court-approved scheme following a period of financial difficulties. On the north-east coast, Dorman Long absorbed Bolckow Vaughan following the failure of the latter's claim for £600,000 against the Department of Inland Revenue in respect of money spent on wartime expansion at the instigation of the government. Two other mergers involved David Colville & Sons and James Dunlop and Company in Scotland and Thomas Firth and John Brown in Sheffield.

Financial reorganization was one weapon forged to meet this situation, protection was the other. In 1917, a Departmental Committee of the Board of Trade had recommended protection for the steel industry in the immediate postwar years, but this was never implemented. The Safeguarding of Industries Act, passed by the Lloyd George Coalition Government in August 1921, provided that duties of up to 33⅓ per cent *ad valorem* could be imposed where dumping was proved, but the Baldwin Government of 1924-9 was reluctant to deal with the problems of the steel industry in isolation. The situation was further complicated after 1926 when, following the expiry of the provisions of the Versailles Treaty regulating Franco-German trade, steel imports into Britain from the continent of Europe increased rapidly. By 1931, much of the foreign steel coming into Britain was being sold at prices below its local cost of production.

The political and financial upheaval of August 1931, which resulted in the establishment of a National Government and the general election in October of that year, made the imposition of protective tariffs for steel a real political possibility. The Import Duties Act of February 1932 provided for a general duty of 10 per cent *ad valorem* on a wide variety of imported goods and created the Import Duties Advisory Committee (I.D.A.C.) to advise the Treasury on the imposition of additional tariffs where these were judged to be necessary. In April 1932, the I.D.A.C. recommended a 20 per cent duty on railway materials, tubes, wire and other highly finished steel goods and $33\frac{1}{3}$ per cent on all semi-finished steel, heavy and light sections and all forms of plates and sheets. This latter rate was imposed initially for three months, but, as this did not immediately stem the flow of continental imports, it was extended till October 1934, on condition that the industry positively set about to reorganize itself. The I.D.A.C. held that the strengthening of the organization of the industry was a pre-requisite to plant modernization and the achievement of an effective agreement with the European steel producers.

The National Federation of Iron and Steel Manufacturers, established in 1918, had made little effective contribution to the formation of a policy for the industry to meet the economic stresses of the 1920's, largely because of a conflict of interests within its ranks. The heavy steel makers favoured protection, while the users of semi-finished steels found it commercially advantageous to be able to purchase cheap continental imports. Moreover, there existed within the Federation some fifty different product associations, often overlapping in products and function, and with some not including all the companies operating in particular areas. Some were primarily price fixing bodies, whereas others were mainly concerned with labour relations. This was the situation which confronted the National Committee, set up by the I.D.A.C. in June 1932, to prepare a reorganization scheme for the industry with the primary aim of creating a body which could take effective and binding decisions on the industry's behalf.

The National Committee's proposed scheme provided for a set of regrouped product associations which would each play a part in framing the Federation's policy and would have adequate means of attracting and controlling members through a system of production quotas and price rebates. These associations were to be affiliated with a strong central body able to exercise the requisite degree of control over production, to ensure concentration in the more efficient plants and to provide a sufficient profit margin to enable them to expand at the expense of obsolete and redundant plants which were to be closed down. The effectiveness of the central organization was to be assured by the support of all the major steel producing companies and financed by a strong central fund created from levies on production.

This proposal was, however, too centralized to meet the wishes of a traditionally independent industry. When the constitution of the new British Iron and Steel Federation (B.I.S.F.) was adopted in April 1934, by 90 votes to 24 with 79 abstentions, there were, in fact, only company members, but with provision for the affiliation of associations on mutually agreed terms. Many of these associations affiliated in later years and today there are ten member conferences of the Federation comprising, in total, some thirty individual associations together with four affiliated associations, which are not members of any conference. By December 1962, there were 441 company members within the ten conferences with some companies holding membership in several.

Another critical provision of the 1934 constitution was that there be an independent chairman in the sense that he had no business connection with any steel company or branch of the industry. This post was occupied by Sir Andrew Rae Duncan from 1934 until 1952, except for the years 1940–5, when he was a minister in the Churchill Government. Given the desire of the I.D.A.C. for a central authoritative voice of the steel industry and the background from which the B.I.S.F. emerged, the qualities of the chairman were of paramount importance. Although no biography of Duncan, apart from a B.I.S.F. memoir, has yet been published, it is evident that he enjoyed a high reputation as a negotiator over a broad area which included the industrialists, the trade unions, the city and the government. Undoubtedly, Duncan was a major influence in shaping the B.I.S.F. in the years immediately following its formation.

The three tasks confronting the Federation in 1934 were:

(i) to make the newly formed organization work in an effective and integrated manner;

(ii) to seek, with government approval, a long term agreement with the European Steel Cartel, in order to bring under control the concerted invasions of the British home market;

(iii) to secure the physical modernization of the industry and to encourage completion of its financial reorganization.

The first of these tasks was pursued by careful and often prolonged negotiations aimed at reshaping and regrouping the product associations and bringing them into the Federation. The second, and perhaps the most pressing, followed a number of clearly defined steps. The first of these was to persuade the I.D.A.C. to remove the time limit on the iron and steel duties of 20 and $33\frac{1}{3}$ per cent which were due to expire in October 1934. The earliest attempt to negotiate an agreement on imports with the European Steel Cartel failed and it was not until the effective tariffs were increased to 50 per cent in March 1935, that provisional agreement was reached. By the end of July 1935, the Federation and the Cartel had come to terms under which imports into Britain were limited to 670,000 tons in the first year, and 525,000 tons in subsequent years, these tonnages being admitted at a reduced rate of 20 per cent *ad valorem*. The entire quotas were to be purchased on the Federation's behalf by its commercial subsidiary, the British Iron and Steel Corporation Limited (B.I.S.C.), created in October 1935.

Shortly after this agreement was negotiated, home demand rose rapidly and to meet the boom conditions of 1937 the duties on licensed steel imports were for a time reduced to $2\frac{1}{2}$ per cent and on unlicensed imports to $12\frac{1}{2}$ per cent. In 1938, however, the boom appeared to have passed, and the duties were restored to their former levels. By 1939 home production stood at 13·22 million tons with total imports of 1·52 and exports of 1·73 million tons.

The third task of securing physical modernization, financial reorganization and production rationalization was one which could only be achieved by persuasion, since the companies involved were responsible for raising their own capital. Despite this, there grew in the years 1936–9, a practice under which companies contemplating major new programmes submitted them for scrutiny by a high level B.I.S.F. *ad hoc* committee. Though this arrangement included no provision for a veto, it created an important route of communication between firms, a possible venue for informal governmental influence, and cannot have been other than helpful in a regulatory sense when outside finance was being sought.

Between 1932 and 1939, £50 million (at current prices) was spent on modernization and the same amount of obsolete plant was written off. Three new schemes, in which the B.I.S.F. did not, however, play any direct role, included the Stewarts and Lloyds' iron, steel and tube making plant at Corby, Northamptonshire, begun early in 1933, the Lancashire Steel Corporation's plant at Irlam on the Manchester Ship Canal, and the rebuilding in 1934 of the Guest Keen Baldwins' Cardiff works. Projects which did come under the scrutiny of the B.I.S.F. committee included The United Steel Companies' new blast and open hearth furnaces at Appleby-Frodingham, the rebuilding and linking of Colvilles' Clydebridge Steel Works and the Clyde Iron Works, and the construction of hot strip mills by Richard Thomas & Company at Ebbw Vale and by John Summers & Sons at Shotton.

The role of the B.I.S.F. at this time in influencing capital investment is well illustrated by the Ebbw Vale development. This was originally conceived by Sir William Firth, chairman of Richard Thomas & Company. His attempts to induce his main rivals, Baldwins, to join the project having failed, he then proposed to proceed alone to construct a fully integrated mill at the Redbourn works of Richard Thomas & Company at Scunthorpe in Lincolnshire. Local and central government pressures resulted in this decision being modified in favour of the Ebbw Vale site where the existing works had been closed since 1929. Work commenced on the South Wales site in 1935, but by 1937 financial difficulties had ensued and the Bank of England, the Federation and Sir Andrew Duncan were jointly involved in the negotiations which led to the provision of financial support.

The interaction between the B.I.S.F. and the Government is shown by the ill-fated Jarrow-on-Tyne development first mooted in 1934. A financial syndicate proposed to erect a plant at Jarrow to make 500,000 tons of pig iron and 400,000 tons of basic Bessemer steel per annum. The Federation, on being consulted, endorsed the scheme on social grounds since 80 per cent of Jarrow's population were unemployed at the time. But the existing north-east coast firms mainly located on the Tees, who were themselves planning to modernize, felt unable to participate and without their backing the scheme fell through. This sparked off the now famous Jarrow march and angry parliamentary debates, with the result that in July 1936 the President of the Board of Trade requested the I.D.A.C. to make a general enquiry into the iron and steel industry. The report of this enquiry, published in June 1937, declared the attitude of the north-east coast firms to be justified; it also made a number of recommendations of considerable long term consequence in the area of public supervision of the British iron and steel industry.

This report pointed to a need for permanent machinery, for example, a wholly independent body acting in the public interest, to ensure that future developments took regard of social considerations; it did not, however, agree that the supervising authority be empowered to promote compulsory mergers. The report also advocated long term forecasting of demand and capacity trends and the adoption of a policy which provided a small margin of reserve capacity to meet possible sudden increases in demand. The pattern suggested or implied for the industry was one involving wide co-operation, self discipline through the B.I.S.F. with quasi-government supervision by a body such as the I.D.A.C. or other statutory authority created for the purpose. This was to be an 'experimental' arrangement designed 'so that there may be a full and fair trial of the possibility of combining individual responsibility and initiative on one hand with co-ordinated and co-operative action and full recognition of the overriding importance of the national interest on the other'.

The I.D.A.C. also played a part in establishing an ordered price structure. Within the B.I.S.F., prices were formally a matter for particular sectional associations and not one for the Executive Committee. But the Committee in association with the I.D.A.C. was aiming at a policy of large unit outputs at low prices. The procedure which ultimately emerged under Duncan's guidance was that, where there was a proposal to increase prices, an investigation of production costs was first to be made by a chartered accountant nominated by the Federation. If his report indicated a *prima facie* case for an increase, the Federation advised the I.D.A.C. accordingly requesting that this body give its approval. In addition, the Federation, at the instigation of the I.D.A.C., also undertook a comprehensive review of the relationship between the prices of the main iron and steel products. After some persuasion of the heavy steel makers by Duncan, such a cost survey was undertaken and was to be forerunner of others which subsequently became the basis for price determinations in the industry. These surveys necessarily took account of capital costs, usually in terms of the two most modern plants in the industry. This uniform cost system, worked out initially in 1935, came into universal use throughout the industry and by the outbreak of the Second World War had become formalized into a system of quarterly cost returns on which price determinations could be made. In addition, to meet abnormal peaks in demand for foreign pig iron and scrap, a 'spread-over' levy was introduced at this time, in order to even out the losses accruing from abnormal purchases of raw materials at times of world shortage.

These various joint activities of the I.D.A.C. and the B.I.S.F., the programme of financial reconstruction and the recovery in world trade enabled the British iron and steel industry to attain a reasonably stable and potentially profitable position by 1939. Dividend payments, which had ceased in the 1920's, were resumed, steel production had reached 13·2 million tons in 1939 and in the period, 1937–9, capital expenditure was running at an average rate of £13 million per annum, compared with about £2 million per annum in the years, 1929–33. Moreover, the industry's central organization had been skilfully reshaped and greatly strengthened.

The years, 1939–45, were not in any sense a period of expansion or innovation. The industry came under the direct control of the Ministry of Supply on 1 September 1939, and from then on the basic aim was to sustain output at around 13 million tons per annum. There were, of course, major changes in product distributions, an enhanced demand for special alloy steels largely met in the Sheffield area, and a host of problems in pig iron and crude steel production arising from the uncertain and more variable raw material supplies of ore, scrap and other ingredients occasioned by wartime conditions especially at sea, and the quantitative and qualitative changes in the availability of labour.

No new steelworks were built in Britain during the war, though four basic open hearth furnaces at the London Midland Scottish Railway Company's steelworks at Crewe which had been shut down in 1932 were brought into production. The balance of steel demand over and above home production was met by imports largely from the United States. Initially, these were handled by B.I.S.C. in the U.S. until the Lend-Lease Acts came into force in March 1941, when this task was assumed by the U.S. government. The pricing arrangements, funds, levies and adjustments which were an inevitable part of wartime conditions need not concern us as they had no direct bearing on decisions to build new or replacement facilities.

The Ministry of Supply Iron and Steel Control Organization was largely staffed at the senior levels by men seconded from the steel companies and from the B.I.S.F. One result of this arrangement was that there was, throughout this period, a high level of co-operation within the industry and, though the period must have been difficult and frustrating in the extreme, it cannot be assessed as one of sterility born of remote central control, or one in which the direction of the industry basically altered in character.

The impact of the war differed widely in the various participating steelmaking countries. While steel production remained fairly steady in Britain (12 million tons in 1943), German, Japanese and French production had fallen to very low levels by 1945. The U.S. had during the war vastly increased its production from 47 million tons in 1939 to 71 million tons in 1945. It might have been expected that the U.S. would have had available a considerable surplus for export in the immediate postwar years, but because of the rapid switch to peace time activities and an almost insatiable demand for consumer durable steel goods in that country, this situation did not eventuate. In consequence, the immediate post war period was one in which on the world scene steel demand exceeded supply.

Two of the features of the British postwar economy, the export drive and the capital investment programme, both made extensive demands on steel, with the result that in the period, 1945–50, demand exceeded supply. In 1950, when it appeared that the gap might be bridged, the Korean war intervened creating fresh demands. One result was that the official control of steel distribution continued, apart from a short break, until 1953. There was a major change in the consumption pattern in these years, with high growth rates in the types of steel most used in the motor, engineering and construction industries.

Unlike the situation after the First World War, the government and the industry were in a much superior position consciously to organize to meet the problems of the postwar period. As early as 1943, the Postwar Reconstruction Committee of the B.I.S.F. was asked to assess the requirements of steel in this period and the steps that should be taken to meet them. The suggested demand figure was 20 per cent above the highest level reached before the war, presumably amounting to 16–17 million tons. This preliminary exercise was subsequently replaced by a series of four Development Plans for the periods, 1945–52–3, 1952–7–8, 1957–62 and 1961–5. These plans each began with attempts to assess the future demands for steel with the results summarized in Table 17.

Table 17: Estimated demand for steel – postwar development plans

Plan	Period	Estimated demand at end of period (million tons per annum)
1	1945–52–3	16·0 (raised to 18 million in 1948)
2	1952–7–8	22·5 (Iron and Steel Board) 21·0 (B.I.S.F.)
3	1957–62	29·0 (Iron and Steel Board) 28·0 (B.I.S.F.)
4	1961–5	29·0 (Iron and Steel Board)

To meet these proposals, considerable expansion and replacement of steelmaking plant was required. However, during the immediate postwar years, there was considerable inflation in capital costs of new plant with the result that the replacement of obsolete facilities was less attractive financially and proceeded at a rather slow rate. At least until about 1960, the emphasis in steelmaking was directed primarily towards the replacement of existing open hearth furnaces with more modern, larger furnaces without a fundamental change in technology. The outputs per furnace[1] shown in Table 18 illustrate this point, although some of the growth may be attributable to improved operating practice on existing furnaces in their original or refurbished forms without significant change in physical size. The higher output rates shown after 1960 are primarily due to the increased use of oxygen. Where new works were established, these were often located adjacent to or near existing works for justifiable reasons, especially when judged on short term criteria.

Table 18: Output per open hearth furnace in United Kingdom
(in thousand tons)

Year	Output	Year	Output	Year	Output
1946	36·8	1952	50·4	1958	71·2
1947	39·1	1953	52·6	1959	74·4
1948	41·4	1954	58·0	1960	77·5
1949	43·6	1955	60·2	1961	78·9
1950	45·8	1956	61·1	1962	89·6
1951	46·8	1957	64·7	1963	95·1
				1964	98·0

It has been publicly suggested that the British steel industry was slow to innovate in the period 1945–60. In comparison with Japan

[1] B. S. Kelling and A. E. G. Wright, *The Development of the Modern British Steel Industry*, Longmans, London, 1964, p. 121.

and Western Germany, this may be formally true in the area of steel-making itself, but there were extenuating circumstances. Initially, this was a period of readjustment after the war rather than one of gross physical reconstruction; demand continued to exceed supply at least until the middle 1950's; the changing product distribution tended to focus new developments in the steel processing and finishing areas or in remelting and manufacture of special steels largely using electric furnaces. Furthermore, as a nation Britain was devoting considerable capital resources to the service and fuel industries, notably electricity supply and coal. In spite of other capital demands of high priority, the total fixed capital expenditure in the U.K. iron and steel industry in these years was not inconsiderable, as the data set out in Table 19 reveal. Of somewhat greater relevance is the proportion of these sums devoted to steelmaking and melting facilities. In the period 1958–63 inclusive, the capital sum in actual terms devoted to this purpose amounted[1] to £130 million.

Table 19: Fixed capital expenditure by U.K. iron and steel industry[2]
(£ million)

Year	At current prices	At 1963 prices	Year	At current prices	At 1963 prices
1946	5	12	1955	58	78
1947	20	44	1956	75	93
1948	30	60	1957	95	110
1949	40	78	1958	105	120
1950	45	83	1959	99	112
1951	50	83	1960	146	159
1952	45	69	1961	199	207
1953	49	73	1962	170	173
1954	52	75	1963	77	77
			1964	75	75

Overlying these developments, a number of organizational changes was taking place within the industry itself and, at the same time, its ownership and direction had become a major political issue. Within the industry, further amalgamations took place and there was a number of joint development enterprises. These need not concern us in detail, except to note that they were a continuation of the rationalization that had begun before the war and were directed either to unifying production in a particular area in order to obtain the advantages of larger scale, or were aimed at securing increased integration, either forward towards rolling and fabrication, or backwards to secure billet supplies.

[1] Mr A. H. Wood, B.I.S.F., Private Communication, 1/3/65.
[2] Kelling and Wright, op. cit., p. 129.

The British Iron and Steel Federation adopted in March 1945 a new constitution which departed in particular ways from the earlier one of 1934. The original instrument, while providing for membership through affiliated associations, allowed direct membership of companies provided they were members of appropriate associations. The companies appointed representatives regionally to the Council on which the associations were also directly represented. In the new constitution, direct company membership was abolished; the affairs of the industry were reorganized into not more than twelve conferences, requiring, in effect, groups of associations in related areas to combine to form conferences; the central body was to comprise the chief executives of the various sections of the industry.

The reorganization was aimed at increasing the power of the conferences and associations relative to the central body. The companies, though not directly represented as hitherto, were the effective operational units in the industry, and there was preserved in the new constitution a nice balance of power between the companies, the associations, the conferences and the central body. The external conditions which were soon to develop, notably those related to nationalization, the establishment of the European Coal and Steel Community (E.C.S.C.) and subsequently of the European Common Market, denationalization, the establishment of the Iron and Steel Board and then a further proposal for nationalization all tended to demand a greater degree of central direction and control than the 1945 Constitution aimed to provide. A further change in the constitution of the Federation was therefore made in 1954, following the appointment of the supervisory Iron and Steel Board referred to below.

We now turn to the question of the relationship between the State and the iron and steel industry which developed in the postwar era. The public ownership by nationalization of the iron and steel industry has so many undertones and overtones in the past, present and future, that it cannot be ignored as a factor affecting innovation in the industry. It is not, however, the purpose here to argue a case for or against this proposal, but rather to set down in outline the main facts as they occurred and to try to assess the possible effects of these events as a whole on the industry's approach to technical innovation.

At the end of the war, the Iron and Steel Control created under the Ministry of Supply continued to operate essentially unchanged except for the progressive return of senior personnel to individual companies and to the B.I.S.F. In April 1946, the Labour Government announced its intention to prepare legislation to provide for a large measure of public ownership of the iron and steel industry and, in the interim, appointed the first Iron and Steel Board with the following functions:

(i) To review and supervise programmes of development needed for modernization of the iron and steel industry and to watch over approved schemes in such programmes;

(ii) To supervise the industry as necessary in current matters, including the provision of raw material requirements, and the administration, under power delegated by the Minister, of such continued direct control as may be required over production, distribution and import of iron and steel products;

(iii) To advise on general price policy for the industry and on fixing prices for controlled products.

The Board thus had advisory and not statutory powers. The areas outlined were ones which were generally similar to those falling within the experience and practice of the industry, the B.I.S.F. and the prewar I.D.A.C., and there were few apparent difficulties in continuing along these familiar lines. The major development of a new strip mill proposed by the newly formed Steel Company of Wales proceeded without hindrance and other proposals in the first development plan were, after perusal, implemented. The position on ore and scrap was already well co-ordinated and organized under the B.I.S.F. trading subsidiaries. On the price question, the first Iron and Steel Board modified the practice originally employed in the prewar years by the I.D.A.C., its nominal predecessor.

The price fixing system of the I.D.A.C. had been based on industry returns of current costs of production plus a fixed standard margin to cover depreciation and profit. In determining this margin, the I.D.A.C. used the capital costs of the then two most modern plants (Cardiff and Corby) and the average production costs were calculated after exclusion of the highest cost producers. During and after the war, the Ministry of Supply had used the full average costs of all the producers and this practice was continued by the Board, but with the proviso that the prewar practice should be restored when supply and demand were more in balance. The Board also introduced revised ways of calculating the standard margin. For the depreciation component, it fixed a rate based on the average written-down replacement cost of the fixed assets of all the works concerned, while for the profit component it fixed a rate of return on the capital employed based on the written-down historical cost of these fixed assets.

The Iron and Steel Bill was placed before parliament in the autumn of 1948 and received the Royal Assent in November 1949. In its passage through the House of Lords, it became tangled with provisions of a new Parliament Bill and, as a result, the earliest date on which the industry could be transferred to public ownership was 1 January 1951, a date which necessarily fell after the next general election. Briefly, the bill provided that the ownership of companies producing or hot rolling annually more than 20,000 tons of steel or producing 20,000 tons of pig iron or 50,000 tons of iron ore would be acquired by the State. An exception was made in the case of companies engaged in the manufacture of motor vehicles, to wit the Ford Motor Company, which owned a blast furnace and rolling mill.

The election of February 1950 returned the Labour Government, but with its majority reduced to six. It decided to proceed with steel nationalization and, accordingly, set about creating the Iron and Steel Corporation of Great Britain to take over the assets and shares of the 94 scheduled companies and 100 subsidiaries which came within the provisions of the act. The Corporation took office in October 1950, and in February 1951 the shares were vested in the Corporation. The 350 companies which did not come within the provisions of the act remained responsible for about 10 per cent of steel production, often in the form of products for which they accounted for as much as 80 per cent. These companies continued as members of the B.I.S.F.

The actual period of public ownership was too short for any substantial reorganization of the industry to have been implemented, but a number of problems came into prominence, not the least of which was the dispute over the Corporation's claim that it should take over the commercial subsidiaries of B.I.S.F., namely the British Iron and Steel Corporation (B.I.S.C.) and the British Iron and Steel Corporation (Ore) (B.I.S.C. (Ore)) which were concerned with buying scrap and ore, respectively, on behalf of the industry as a whole.

Eight months after vesting day, in October 1951, the Labour Government was defeated in a general election and in the following month the new government declared its intention to denationalize the industry. A new Iron and Steel Bill was introduced into parliament in November 1952, and received the Royal Assent in May 1953. In the meantime, a stand-still order had been issued to the Iron and Steel Corporation of Great Britain instructing it not to make any changes in the financial structure and management of the industry without first consulting the Minister of Supply.

The Iron and Steel Act of 1953 repealed the Act of 1949, dissolved the Iron and Steel Corporation of Great Britain, created the Iron and Steel Holding and Realization Agency (I.S.H.R.A.) for the purpose of returning to private ownership the nationalized undertakings and established the Iron and Steel Board, 'to exercise a general supervision over the iron and steel industry . . . with a view to promoting the efficient, economic and adequate supply under competitive conditions of iron and steel products'. In seeking to combine private enterprise with public accountability, the act laid specific responsibilities on the Board and prescribed some powers to enable it to discharge them. In brief, the Board was to keep under review productive capacity, prices, supply and distribution of raw materials and fuel, promote training, education and safety, health and welfare of employees and provide for joint consultation between management and workers on matters other than the terms and conditions of employment. The responsibilities in this list of relevance in our present context are now considered briefly in turn.

The Board assumed responsibility for prices in December 1953. The act required it to keep under review prices of certain defined iron and steel products and permitted (not required) it to determine maximum prices at which these products (except castings and forgings) could be sold in the United Kingdom by the producers. In its first comprehensive price review in 1954, maximum prices were based on: (i) the average cost of production of each product by the more efficient producers; plus (ii) a margin for depreciation and obsolescence of plant at replacement cost; plus (iii) a margin for profit based on historical capital employed. The Board soon recognized, however, that such a system might not reflect quickly enough the possible reduction of costs resulting from new technical developments, and might not offer enough incentive to expansion by providing sufficient return on new plant at a time of rising costs. To foster within its pricing policy technical modernization and innovation and to discourage the use of obsolete plant, a new principle was subsequently adopted. Something like an ideal plant was created on paper and its costs worked out on the basis of the most advanced type of plant in commercial operation in the United Kingdom. Maximum prices were then fixed on the basis of this hypothetical plant to cover operating costs with margins for depreciation and profit.

12

Another aspect of pricing is concerned with the supply and cost of raw materials, for if these were left in a wholly uncontrolled state, the U.K. costs would, in respect of imported raw material items, show all the variations of the world supply position. This problem had long been recognized by the industry, itself, and had been effectively resolved by centralizing the procurement of these supplies through the commercial subsidiaries of the B.I.S.F., namely, the B.I.S.C. (Ore) and the B.I.S.C. Associated in part with these arrangements, an Industry Fund had also been established and administered by the B.I.S.F. Most producers were signatories to the Industry Fund Agreement, and the Iron and Steel Board, by recognizing these levies imposed to support the fund as a component in the costs of production, effectively endorsed the industry's own efforts to regularize its central services.

The pricing policy and procedures of the Board have not been without controversy. The prices set are maximum, legal prices determined on the basis of new efficient plant costs. They are the prices which the Board expects will, in general, be the appropriate selling prices and are, in fact, those which the Federation recommends to Associations which in turn recommend to member companies. A recent judgment by the Restrictive Practices Court in the Heavy Steel Association's case would appear to mean that an agreement within an Association to fix the price at this maximum, permissible level, or at any other, is illegal. Even without an agreement, there will be little incentive on the part of individual companies to sell below prices set from outside the industry by an independent statutory authority. Such competition as exists is likely to be in terms of service, delivery times, goodwill and other less tangible aspects of trading.

Export prices are not subject to the Board's maximum price determinations and, since the total volume of world exports is rather less than 10 per cent of total production, surplus home production will often be disposed of on the export market at whatever price it will fetch. In these circumstances, the export market in steel has been very unstable and intergovernmental moves are currently being pursued to try to bring some stability in this area. Its relevance to our present study is not so much related to the question of British exports, but rather to the vulnerability of the home industry to dumping in periods of excess world supply; this occurred in the 1961-3 recession.

A second area in which the Board was required to act concerns estimates of future demand. This is a complex task as it is affected by the consumption patterns in a vast range of industries, the incidence of technological change within these industries and the changing patterns of local production and consumption and in traditional markets abroad. The success or failure of the Board's complementary task of supervising the expansion and modernization of plant rests heavily on the accuracy of these long term assessments of demand.

The powers of the Iron and Steel Board in respect of new developments, expansion and modernization and, in particular, the way in which these powers are exercised are specially important. All major schemes of development or replacement (at present defined as those costing in excess of £100,000) have to be submitted to the Board for its consent. The Board normally looks at any proposal from several standpoints, for example, whether there will be a demand for the product in question, whether the raw materials are available, whether the capital costs are reasonable, whether the proposal embodies efficient, up-to-date practices and techniques and whether the estimates of profitability indicate that the proposal is likely to bring about a reduction in costs immediately or in the future. Nearly all such schemes are subject to two other analyses outside the Board, the first in the submitting company which has to find the necessary capital, and the other, a voluntary one, by the Development Advisory Committee of the B.I.S.F. The latter may offer or be asked for advice on particular projects by the Board.

These provisions appear, at least in principle, to place considerable powers in the Board's hands. Consent to a proposed scheme may, however, be refused only if it would seriously prejudice the efficient and economic development of the industry and, in such cases, the company concerned has the right of appeal to the Minister. In practice, the Board has seldom refused consent, but has on occasion had schemes modified as a result of consultation with the company or companies concerned. The Board also has positive powers, in that if the developments it judges necessary are not forthcoming from existing producers, it may so report to the Minister of Power who may make arrangements for a company as his agent or otherwise to provide the necessary facilities, or he may do so himself—whatever that may mean. The Board is not, however, empowered to finance development. In practice, this situation does not appear ever to have arisen.

In 1962 the Board drew attention to the fact that total capacity of the industry was in excess of demand and that the circumstances were opportune for the withdrawal of older, less efficient capacity. The Board has no statutory powers to direct that capacity should be withdrawn. This decision rests with the individual producers themselves, and since withdrawal of obsolete plants will usually result in unemployment, there are pressures on the Board not to push this issue too vigorously.

The operations of the Iron and Steel Board since 1953 have been an interesting experiment in state-industry relationships. Within its terms of reference, the Board appears to have discharged its functions more by using consultation and persuasion than by the vigorous exercise of the full powers vested in it. In so far as the Board has directly influenced the development of the iron and steel industry, the overall result has been minimal and negative rather than excitingly positive, as, perhaps, its sponsors deliberately intended. In its pricing policy and determinations, there is little evidence to suggest that its decisions have unduly sustained obsolete or inefficient plants; in the limit, its methods of calculation should encourage modernization. In its development activities, the Board can in principle see that unnecessary duplication of facilities is avoided, but there have been cases in which the Board has not acted very positively in this regard.

Again, the Board can examine proposals to see that they meet the standards of up-to-date technology, but there is an inbuilt, unstated implication that new proposals should not be too revolutionary. It is a matter of common experience that it is far easier to estimate accurately the capital requirements and operating costs of a new plant that has evolved from previous, well established experience than to achieve like accuracy with a novel process or design. Moreover, many of the savings in the manufacture of steel products are likely to come from an integrated plant with balanced capacity in its several components. Greenfields situations tend to be the exception rather than the rule, and the merits of replacing a particular unit within an integrated works by a new alternative can seldom be judged in terms of simple, isolated, generally applicable criteria. In addition, it is misleading to take too regular a view of capital investment in and depreciation of steelmaking furnaces. Such furnaces and some of their ancillaries are subject to periodic relining and partial or substantial reconstruction resulting in a concomitant reassessment of their capital value and depreciation rate. When this is related to a long total life, often extending over more than thirty years, during which there has been a large escalation in capital costs, some rather arbitrary accounting answers can be obtained.

In general terms, the particular organization of the industry as a whole, in companies, associations, conferences and federation, and the exercise of the Board's powers in the manner outlined above, all tend to point to a situation in which evolutionary innovation and the introduction of locally new but industrially proven techniques are encouraged, but 'blue-sky', revolutionary innovations are discouraged, if not directly, then by the particular system of checks and reviews which has been instituted.

In proposing this conclusion, it is necessary to recognize clearly a number of other facts. Firstly, Britain is a comparatively small steel producer with about 6 per cent of the total world production (in 1963). The second is that, compared with the 1920's and 1930's, the iron and steel industry has since the Second World War enjoyed a considerable period of profitability and of market stability with some degree of public supervision, but without the extensive and detailed apparatus of government control and direction. Possibly the best ultimate test of its performance is to be found in a comparison of steel prices in Britain, U.S. and Europe. Table 20 extracts figures from a more extensive compilation made by the Iron and Steel Board.[1]

From these figures, a comparison for this range of products made by open hearth methods shows that only in Belgium are the domestic prices lower than in the United Kingdom. Not unexpectedly, basic Bessemer steels are generally cheaper than those made in open hearth furnaces, but the ability to use the former process is largely a function of pig iron composition which is directly related to the ore used in iron making. In all the countries mentioned, steelmaking is a mature industrial activity and, in so far as the extensive use of oxygen either in converters or open hearth furnaces results in price reductions, these are likely to be of the same order in all cases. The extent of the potential price reductions arising from a substantial change towards oxygen-using processes seems unlikely to exceed 10 per cent, though information on this point is not available for disclosure.

It is, however, necessary to add some qualifications in respect of the figures cited which are all published prices adjusted uniformly to include delivery to customers. The United Kingdom figures are the maximum prices determined by the Iron and Steel Board, but in

[1] Iron and Steel Board, *Annual Report 1963*, H.M.S.O., pp. 62–3.

Table 20: Steel prices in U.K., U.S.A. and E.C.S.C. at 31 December 1963

Product	Quantity	U.K.	U.S.A.	W. Germany		Belgium		France		Luxembourg	Italy
		O/H	O/H	Basic Bessemer	O/H	Basic Bessemer	O/H	Basic Bessemer	O/H	Basic Bessemer	O/H
Billets Tested 4"	100	33/13/-	40/9/-	33/9/6	37/1/6	29/12/6	33/5/-	31/14/-	35/1/-	32/2/-	36/2/-
Plates Basic quality	50	42/3/-	55/1/-	42/18/-	47/19/-	35/5/6	37/1/6	42/5/6	46/6/6	43/7/-	45/0/6
Heavy sections Basic quality	50	40/3/6	53/17/-	41/2/-	44/17/-	37/0/6	40/13/-	39/8/-	43/5/6	37/12/-	40/13/6
Wire rods Soft basic	50	41/12/-	59/3/-	40/14/-	44/9/6	31/5/6	37/16/6	39/1/-	41/12/6	37/18/6	43/6/-
Hot rolled strip Basic quality 12"×10G	50	42/8/6	51/17/-	42/7/-	46/3/6	41/16/-	48/6/6	40/5/6	44/15/6	36/13/-	41/12/-
Sheets General purpose 6'×3'×20G	50	54/12/-	61/13/-		62/17/-		59/7/6		58/12/-	60/2/6	56/1/-

some European countries the prices published by different producers vary somewhat. In such cases, the *lowest* published price has been quoted. In E.C.S.C. countries, producers are required to publish price lists, but are free to fix their own selling prices which, in practice, may be below the listed figures. In times of weak demand or where marginal selling is resorted to for other reasons, the prices, in general, will be below those officially listed or recommended by central bodies in the industry. These qualifications do not, however, detract from the validity of the statement that British steel prices are fully competitive with Europe and U.S.A.

Finally, it is necessary to say something about some recent political events closely affecting the steel industry in Britain. The Iron and Steel Holding and Realization Agency set up under the Iron and Steel Act of 1953 was given the task of returning to private ownership the companies which had been nationalized. By January 1955, 50 per cent of the steel output had been returned to private hands and by 1957 this figure had risen to 86 per cent. By the end of 1963, I.S.H.R.A. was left with only Richard Thomas and Baldwins Limited. This company has remained in public ownership and has increased in scope somewhat by the purchase of the Whitehead Iron and Steel Co. This purchase was approved largely on the grounds that Whiteheads took more than half the billets produced by Richard Thomas and Baldwins Limited at its Redbourn (Lincolnshire) and Gowerton (West Wales) works.

The current situation has recently been dominated by a decision of the Wilson Labour Government to proceed again with steel nationalization. The White Paper setting out the government's intentions proposes to take into public ownership, in addition to Richard Thomas and Baldwins Limited, thirteen companies, each of which produced in excess of 475,000 tons of crude steel in the twelve months ending 30 June 1964. These fourteen companies operate twenty-two integrated works and control over 90 per cent of the production of iron ore, pig iron, crude carbon steel, heavy steel products, sheet and tin plate. The companies (including their subsidiaries) which are named in the White Paper are as follows:

Colvilles Limited
Consett Iron Co.
Dorman, Long & Co., Limited
English Steel Corporation Limited
G.K.N. Steel Co. Limited

Richard Thomas & Baldwins Limited
Round Oak Steel Works Limited
South Durham Steel and Iron Co. Limited

John Summers & Sons Limited
The Lancashire Steel Corporation
Limited
The Park Gate Iron and Steel Co.
Limited

The Steel Company of Wales
Stewarts and Lloyds Limited
The United Steel Companies
Limited

The remaining 347 limited liability companies in the industry are excluded from the initial proposal, but there is provision to extend its scope if this is thought desirable in the future.

There are yet no firm ideas on the detailed organization of the industry, as this is one of the first tasks listed for the attention of the National Steel Corporation which is to be created. Briefly, the three principal reasons advanced in the White Paper for the proposal are as follows:

(i) the powers of the present Iron and Steel Board are essentially negative;

(ii) the extent of public moneys invested in the steel industry over the past ten to fifteen years and the need in the future for investments in large quanta of a kind not attractive to the private investor;

(iii) the pricing system operated under the Board is monopolistic and not in the public interest.

It is not at this point in time worth considering in any detail the ultimate effects on the industry when this proposal is implemented. The present opposition party has declared its intention, when it is returned to power, of again reversing the process and returning the industry to private ownership. Whatever may be the rights or wrongs of these countervailing attitudes and policies, it can scarcely be maintained that a further period of uncertainty if not upheaval will be conducive to technical progress.

Research in the British iron and steel industry

Any study of innovation in an industry, or in one segment of an industry, properly requires a review of the relevant research effort from the viewpoints of organization, character, extent and manner of prosecution. Under the Iron and Steel Act of 1953, the Iron and Steel Board was required to keep under review the arrangements for the promotion of research relating to the industry, and was given power to require contributions from producers of iron and steel and to make loans or grants-in-aid to promote research relating to the industry. In performance of these functions, the Board published[1] in 1963 a special report on this subject.

Any analysis or review of the research and development activities in any particular industry or in industries generally begins with the difficulty in defining what is to be included in this area. In broad terms, these activities may range from sponsored or unsponsored fundamental research in university-type institutions, through long and short term studies in industrial research laboratories and establishments and in research associations to investigational work directly on a production unit itself with object of improving its performance in one or more particulars, or of obtaining information useful as a basis for some further development. The cost of all or some of these activities may appear separately under the heading, research, in a somewhat ill-defined category described as development, in a joint category of research and development, or they may be absorbed in production costs or overheads and not appear as a distinguishable item in any set of accounts. The particular practice followed varies from industry to industry and country to country and depends, *inter alia*, on the taxation laws, regulations and nomenclature of the country or countries in question and the traditions and practices in the industry. The iron and steel industry is not adequately described as a simple unitized industry, but is composed of companies and groups of companies varying enormously in size, sophistication and range of activities from integrated large scale operators to small iron and steel foundries.

The nature of the principal methods of iron and steelmaking and of subsequent steel processing are such that it is difficult at the present level of scientific and technological knowledge to scale up effectively with any acceptable degree of certainty from laboratory to pilot scale and from pilot to full scale operation. Consequently, there remains for the present a high degree of empiricism in this area with a high value placed on operational experience and an overriding need to try out new ideas, at least at a pilot scale and preferably at a full scale as well, at an early stage in their history. Effective research of this kind is likely to be inconvenient and expensive, with the result that the bulk of the effort in the basic activities of iron and steelmaking tends to be directed to the evolutionary development of existing process-types in preference to quite new or revolutionary processes.

[1] *Research in the Iron and Steel Industry*, Special Report, 1963, Iron and Steel Board, H.M.S.O., London.

We have referred in the last section to the high and growing capital cost of new plant and to the particular feature of periodic regeneration of the asset value in such units as blast furnaces and steelmaking furnaces arising from major relining and rebuilding. These circumstances afford convenient opportunities to include modifications, but are not conducive to changes of a very major kind. The overall result is that a given steelmaking habit, practice and facility once established is likely to persist for a very long time unless quite a major change in total circumstances occurs.

A further point of significance is that there is associated with steelmaking a major investment in ancillary plants and services, for example, those required for materials handling both in and out of the melting shop. The speed and convenience with which furnaces can be charged and tapped may be determined by the limitations imposed by crane capacity, or scrap assembly, or staging facilities, or soaking pits so that quite new steelmaking furnaces will frequently demand a major new investment in facilities of this kind. In Britain, this occasionally presents special problems arising from the somewhat constricted or rather unsuitable sites on which steelworks have in the past been built.

THE ORGANIZATION OF RESEARCH

Depending on the traditions, history, laws and needs in different countries, research in the iron and steel industry involves in varying proportions work carried out by central institutions financed by the industry as a whole, with or without government support, and by companies or groups of companies operating individually. In Britain, the British Iron and Steel Research Association (B.I.S.R.A.) carries out a substantial amount of research of common interest to iron and steel companies, but the largest part of the effort is borne by the companies themselves. B.I.S.R.A. is financed ultimately by company contributions through the British Iron and Steel Federation, and in addition receives a government grant as a research association under the Department of Scientific and Industrial Research[1] (D.S.I.R.). In 1964, its total income was £1,380,539 which included a basic subscription from the industry of £1,061,795, a government grant of £159,269 together with minor amounts from sundry other activities.

---

[1] In 1965 this department was abolished and its several functions taken over by various ministries, the Science Research Council and other similar bodies.

The pattern of research financing during the past decade is summarized in Table 21.

Table 21: Research and development expenditure — British steel industry
(£ millions)

Source	1954–5	1958–9	1962	1963
Companies*	3·0	4·0	6·3	8·0
B.I.S.R.A.†	0·5	0·8	1·1	1·3
Total	3·5	4·8	7·4	9·3

*Includes £0·75 million not within the field of interest of B.I.S.F.
† Includes grant from D.S.I.R.

These figures suggest that in recent years the B.I.S.R.A. expenditure has been about one-seventh of the total. The Iron and Steel Board Special Report states that 'companies in the general branch of the industry spend about three times as much on their own research as is spent by the industry's research association.' This apparent discrepancy probably reflects definition problems with respect to research and development.

In France, there is also a central research association, the Institut de Recherches de la Sidérurgie (I.R.S.I.D.), which is financed by a voluntary levy on sales paid by the steel companies in that country; there is no government grant. Expenditure by I.R.S.I.D. in 1963 was £2·7 million. In Belgium, Luxembourg and the Netherlands there is a joint central organization, the Centre National de Recherches Métallurgiques (C.N.R.M.), which is financed by the steel industries of the Benelux countries and the Belgian government; in 1962, the total C.N.R.M. expenditure was £570,000. In Western Germany, centralized research is organized through the steelmakers' technical association, the Verein Deutscher Eisenhütteleute (V.D.E.). It has power to levy steel producers and provides funds for research at the Max Planck Institute for Iron Research at Düsseldorf. The work in this institute is mainly fundamental in character and involves an expenditure of £450,000 per annum of which V.D.E. provides £270,000, the remainder coming from the provincial government and local iron and steel firms. V.D.E. also financially supports research at the Technische Hochschule at Aachen which, with its 10,000 students, is the largest centre in Germany for training iron and steel technologists.

The central iron and steel institute in Stockholm, Jernkontoret, has no laboratories of its own, but sponsors research in the Metallographic Institute in the University of Stockholm and finances and arranges for collaborative programmes to be carried out in particular steelworks on its behalf. In Italy, there is no co-operative effort, though individual larger companies maintain substantial establishments. In eastern Europe, research institutes roughly comparable in size with B.I.S.R.A. exist and enjoy support from the industry and the government.

Much of the iron and steel research in the U.S.S.R. is carried out in the Central Scientific Institute of Ferrous Metallurgy which maintains a central laboratory in Moscow and an experimental steelworks at Nova Tula, 110 miles from the capital. The design and development of steelworks are handled centrally in the Soviet Union, but individual works are encouraged to undertake research on special problems of their own and in the general area of improving methods of production. A British delegation in 1958 estimated that the total expenditure was about three times that of the corresponding British effort.

In proportion to output, it is probable that the total research effort in the United States is considerably larger than in Britain, but this is a situation which appears to have emerged in the postwar years. There is, in the light of the anti-trust laws of the U.S., no co-operative research association or collaborative effort and it is possible that there is considerable duplication of work in the major steel companies. The U.S. government does, however, make some contribution through the Bureau of Mines which for some years has maintained an experimental blast furnace. The situation in Japan follows the U.S. pattern with substantial company facilities, but no formal central body. The estimated annual expenditure in Japan is £8·5 million, i.e. about 0·85 per cent of turnover.

In addition to the national efforts of the kind described above, there are some international programmes of significance. The High Authority of the European Coal and Steel Community committed £3·6 million in the years, 1953–64, for various research projects of common interest to the Community excluding work on new sources of ore. The International Flame Research Project begun in 1949 involves the E.C.S.C., which contributes 35 per cent of the income, together with Belgium, France, Germany, the Netherlands, U.K. and U.S.A. In 1962, the project's budget amounted to £61,000. B.I.S.R.A. has joined with C.N.R.N. in a project on continuous casting and with the U.S. Steel Corporation on a programme on the mathematical analysis of factors governing the efficiency of blast furnaces.

In summary, it is evident that, except in the special cases of polar, political philosophies in U.S.A. and U.S.S.R., there is a fairly general pattern of a central co-operative research organization operating along with the individual research programmes of separate companies. The distribution of effort between these two contributions appears to be comparable in Britain and Germany; in Belgium and France the central bodies are responsible for a larger proportion than in Britain; in U.S.S.R. the centrally conducted work constitutes an even larger share, whereas in U.S.A. and Japan there is no co-operative programme on the European model.

Various figures and indices have been proposed for assessing the relationship of research effort to the total productive effort. These are subject to considerable uncertainty, especially when the comparisons are made between different industries in the same country or between the same industry in different countries. The Special Report of the Iron and Steel Board considered this matter in some detail and concluded that, in relation to output, the British effort was rather below par, and that in the future the industry should aim at a research expenditure of about 1 per cent of turnover, representing about £350,000 per million ingot tons on current values. Very roughly, this is double the average level of expenditure during the past few years. Quantity is, however, no substitute for quality or selectivity and for our present purpose it is more relevant to look at the extent and pattern of British work in the area of steelmaking.

RESEARCH IN STEELMAKING

In any industrial research situation, whether in an individual company or co-operative or central research organization, there will usually be more possible programmes than there are people and other resources to carry them out. Programmes if rationally chosen are or should be based on some assessment of priorities. In any typical situation in the iron and steel industry, there exists at any one time an enormous range of possibilities from, say, beneficiation of ore at one end to meeting the demands for special alloy steels for rocketry or turbines at the other. Here we are concerned with an assessment of the priority proper to steelmaking in Britain since the war.

Between the two basic areas of ironmaking and steelmaking, the immediate postwar period resulted in considerable emphasis on the former with measurable success. For example, the average consumption of coke per ton of iron made in 1945 was about two and a half times that today, and the average output per furnace is now about two and a half times the level of 1945. The reasons for the concentration of effort in this direction are not hard to find. Basically, they arise from the fact that British home ores are lean by world standards and there is necessarily considerable dependence on imported ore. Improved fuel consumption, higher output rates using prepared burdens of sinter and pellets, higher top pressures (though these are not widely used) and the use of auxiliary fuel by way of oil and tar are likely to offer a better return in money than an equivalent effort in steelmaking. At the end of the war, most steelmaking capacity was in open hearth furnaces which had a high level of flexibility with respect to the types of feed they would accept and the product they could make. The simple answer to increased production, quite apart from improved techniques made possible by a greater uniformity in the quality of pig iron arising from the use of prepared burdens in the blast furnace, lay in larger furnaces of somewhat improved design with shorter melting and refining times.

When oxygen became available in quantity, it is not surprising that British effort was directed to utilizing this agent primarily in open hearth furnaces for combustion and refining. The open hearth is a rather slow process and a 25 per cent gain in output simply by putting sufficient oxygen in the furnace could be a highly profitable exercise requiring minimum additional capital. There was also the question of ultimate product distribution. The postwar years saw a considerable growth in the proportion of steel being utilized in strip and sheet form, and for these purposes open hearth steel did not suffer from the disadvantages of Thomas steel with its high nitrogen content.

There were, therefore, many pressures directed towards improving the production rates of open hearth furnaces by the use of oxygen. Many companies have been involved in this, perhaps the most noteworthy being the early work on roof lances carried out by the Steel Company of Wales, the progressive developments instituted and sustained by Guest Keen Iron and Steel Works at Cardiff and the formal integration of intensive oxygen usage in tilting open hearth furnaces in the Ajax process fully developed by Appleby-Frodingham at Scunthorpe. The incentive for experimentation in top blown converters was greatest at such works as Richard Thomas and Baldwins at Ebbw Vale and at the Shelton Iron and Steel Works at Stoke-on-Trent where there was an apparent need soon to replace existing furnaces of rather small size, or where, as at Consett, an early approach to obtaining greater production from existing open hearth furnaces had been to adopt a practice of duplexing using a Bessemer converter. Oddly enough, top blowing with oxygen-enriched air in Tropenas converters had been used commercially in steel founding as early as 1947 by Catton & Co. of Leeds.

The invention and development of the Fuel-Oxygen-Scrap process by B.I.S.R.A. was directed to meeting a situation which, though not peculiar to Britain, was nevertheless very important there. Electric power is, in general, high in price in the United Kingdom and the case for its use in steelmaking has to be made in economic terms. For special alloy steels which attract a high price and are made in small batches to closely defined specifications, electric steelmaking is acceptable economically and has considerable additional technical advantages. For common steelmaking in electric furnaces, the contributions to total cost of the power consumed is likely to be acceptable only when the process is operated on a large scale with multiple units in order to make more uniform the power demand. This is, in part, the situation that has developed at Steel, Peech and Tozer in Sheffield. As the average size of electric furnaces increases, as it has done markedly in recent years, there emerges a demand for a fairly simple steelmaking process using a total or near total scrap charge which, while similar to electric steelmaking, does not suffer from the cost disadvantages and is not so restricted in the scale and organization necessary to achieve acceptable costs. It was to meet this demand that the F.O.S. process emerged. Whether this process ultimately plays a significant role in steelmaking in Britain is a question which cannot yet be adequately answered, but it is undoubtedly an intelligent and resourceful response to a potential situation in this country.

Finally, it is worth reiterating that in the century, 1860–1960, Britain's position as a world producer of steel has fallen steadily. On the present world situation and on any realistic prediction, the possibilities of vastly expanding the production of common steels in Britain with the avowed object of assailing the export market in this product as such is almost certainly doomed to commercial failure. A far stronger case can be made for placing enhanced effort on the production of lower tonnage specialist items, on new and better designed steel sections, on materials, forms and finishes of high quality and in products and areas of steel usage in which there is a large component of innovatory skill and in which the added value in manufacture is high.

THE APPROACH TO RESEARCH IN THE
IRON AND STEEL INDUSTRY

The ultimate results of industrial research depend not only on the particular level or organization of the material facilities, but also of the number, quality and motivation of those involved in sponsoring, directing, executing and implementing the results of the programmes, and on the climate in which particular projects are carried out. Any enquiry in this area is, by its nature, highly subjective, but the possible influence of these factors cannot be properly set aside on this score.

In the British universities, metallurgy has not enjoyed either the vogue or the status of the traditional sciences of physics, chemistry and biology. Though it often initially developed out of chemistry, the research contributions from the universities to physical metallurgy— perhaps, more strictly, the physics of metals—since the war have been outstanding. These have come not only from Departments of Metallurgy, but also from Departments of Physics. Metallurgy treated as an off-shoot of chemistry, physics or engineering no longer represents a realistic appraisal of its organization; it is now formally and widely recognized as a discipline in its own right with well defined areas of physical metallurgy, chemical metallurgy and metallurgical engineering. Currently, it is tending to move towards becoming the core of the rapidly evolving area of materials science,[1] now represented by separate chairs in several universities in the United Kingdom. The direct impact of this considerable research effort in physical metallurgy on steelmaking has been small, though this is not so in the area of special steels deliberately designed to have particular mechanical, electrical or magnetic properties.

University research in chemical metallurgy has been rather more restricted. Thermodynamics in relation to metallurgy has received some much needed attention, though the total effort and the tradition of work in this area in the United Kingdom scarcely measures up to that in Germany and the United States. The study of kinetic processes, especially slag-metal reactions, is a peculiarly difficult field and has not advanced to a stage at which it makes more than a nominal impact on industrial steelmaking. The problem in part is that the mass and heat transfer conditions which exist in a steelmaking furnace and the interrelated reactions which are the

---

[1] See for example R. W. Cahn, *Discovery*, *26*, 41, July 1965.

basis of steel refining are notoriously difficult to recreate on a scale and under a degree of control that makes work on a laboratory scale in this area directly meaningful or applicable. Furthermore, a substantial part of the research effort in chemical metallurgy, whether in universities or research establishments in Britain, has tended to be concentrated in the non-traditional areas related to metals such as titanium and to those materials and systems closely related to the atomic energy programme. It is in the nature of universities, especially in Britain, that there is little interest in those areas which make or could make the most significant contribution to traditional, industrial processes.

In the neighbouring areas of fuel technology and chemical engineering, generally similar considerations apply, petrochemical processes, for example, appearing far more exciting than those of steelmaking. In addition, the comparatively poor mineral resources in Britain have not led to vital university schools of applied geology which, in some parts of the world, spill over into areas of ore beneficiation and burden preparation. Engineering generally in Britain, except in certain aspects of the electrical field, has had a mixed history in the British universities. With a growing number of student places currently unfilled, some effort is at last being made to attract an increasing number of young men with a physics-mathematics bent into this faculty, though this move has to contend with contrary currents in English education manifested in many schools.

In summary, it appears that, apart from special steels and steel alloys, the situation between the iron and steel industry and the universities is one of severe detachment. The situation is not, however, without the prospects of change arising in part from the emergence of newly organized areas such as materials science, and the possible modification of traditional attitudes which may emerge from the current programme of bringing into one domain the older universities and the newer Colleges of Advanced Technology which have frequently much closer ties with industry. Let these comments not be construed as advocating that the universities should direct their attention largely to existing industrial problems, but rather to pose the quite crucial problem of how the universities in Britain are, on the one hand, to retain a necessary degree of detachment, independence and useful isolation, and, on the other, come to recognize more effectively their responsibility to the community which nurtures them.

Of the industry research associations common in many fields in the

United Kingdom, few are more widely and favourably recognized internationally than B.I.S.R.A. Since its inception in its present form nearly twenty years ago, it has grown greatly in stature and influence. Its research record broadly based and relevant has on most criteria of judgment been successful. No narrow concept of laboratory research has been followed and a substantial amount of process research on the plants of individual companies has been carried out. Considerable attention has been directed to the area of operational research, to the invention of techniques and instruments of measurement and control and to computer usage and operation in the industry. Fuels, as well as other raw materials, have received due attention. A further striking feature is the extent to which members of staff of the Association after a few years have moved to posts in the iron and steel industry itself, and the generally good relationships that appear to have been established at several levels between the Association and the industry, not only in the formal terms of multiple technical committees, but on a personal level as well. There is, however, a common, somewhat muted criticism of B.I.S.R.A. heard within the companies, though this is possibly more pronounced at some levels than at others. The burden of this criticism is that the Association's programme and approach are too 'academic', a statement which in most individual cases is to be construed as meaning not relevant to the production problem currently confronting the particular critic.

This is not a problem peculiar either to the iron and steel industry or to B.I.S.R.A., but occurs to a greater or less degree in the relations of any co-operative research organization working on behalf of an industry. In large measure it is a function of the nature of research associations and one of the limitations of this organizational form of industrial research. Despite this, B.I.S.R.A. has probably gone further than almost any other research association to minimize the effects of this problem; its geographical decentralization, the choice of its programme and its operational spirit are features much to be commended.

Some of the steelmaking companies, either individually or in terms of their group structure, conduct research and development as a physically distinguishable activity, often in separate laboratories located some distance from one or other of their works. These, too, suffer from some of the disadvantages of B.I.S.R.A. in that once a research idea reaches the stage at which it needs to be tried out on a plant, it begins to interact with production schedules. Again, this is not a problem peculiar to the iron and steel industry, but is perhaps a little more strongly in evidence there than in, say, the chemical industry. All who work in any formal organization are subject to judgments of performance of one kind or another. The criteria are, however, quite varied. The performance of a production man will, in general, be judged in terms of his ability to meet production schedules to time, to quality and to cost. A research and development officer, though probably not assessed in terms as mathematically concrete, is nevertheless judged in the light of his performance. What the former conceives as an appropriate approach to meet his criteria of performance, his interacting colleague in a separate but cognate activity might find an inhibiting or discounting factor in meeting his own performance target.

One result of this situation is that, not infrequently, the research and development department directs its programme to either or both ends of the productive operation and not much towards the centre. For example, it is not uncommon to find research programmes heavily weighted towards raw materials and fuel at one end and towards the final product, in such forms as heat treatment, quality assessment and so on, at the other. In this way, the interaction of research and development with the central productive function tends to be minimized.

Another important question concerns the kind of people who are involved in detail with the choice of new processes. Arising, in part from the situation noted above and the traditional practices in the industry, the pattern which has commonly but not universally developed in Britain is briefly as follows: In respect of steelmaking capacity, the decision in principle to expand, contract, reconstruct or replace an existing facility is made in the first instance on the basis of straightforward statistics and estimates of existing and required capacity, in relation to costs, market demand, and other commercial considerations. The working team responsible for converting this proposal into concrete terms will often comprise three or four people including the works manager, the chief engineer and the steelworks' superintendent and possibly a person having a central technical function, such as the chief metallurgist. The general practice is then to visit other steel plants, to review the existing possibilities and to produce a proposal which may be crudely described as the best existing practice 'plus a bit'. The last phrase often incorporates a multitude of small items or ideas of an adaptive character suggested by local experience.

One aspect of this practice is that the selection of the process is made by a group which is almost universally oriented towards production. This has two dominant effects: (i) the new process to be introduced can normally be brought into production rapidly, as its design and construction tend to include the quintessence of operating experience; and (ii) the choice of process is likely to be evolutionary rather than revolutionary and, if it is locally quite new, its assessment will usually be in terms of trials done or observed on a similar plant already in existence and operating on a commercial scale, if necessary in another country.

Such an approach is feasible because there exists within the iron and steel industry in the western world, and extending on occasions beyond the iron curtain, a remarkable degree of freedom in the exchange of visits and of information in a manner almost unknown in many heavily science-oriented industries. The information exchanged is not simply direct know-how, but often includes information which permits a good appraisal of local costs to be made if a similar plant were to be established. The total effect of all these factors is to weight the scales in favour of the steady evolution in steelmaking processes and to discount the prospects of revolutionary change.

Oxygen steelmaking in Britain

In this section, the current (1964) situation in respect of intensive oxygen steelmaking in Britain is reviewed and an attempt made to set out the reasons underlying the decisions of individual companies and works. The information and its assessment rest mainly on two sources: (i) published data, particularly in official, regular and special reports of the Iron and Steel Board; and (ii) information elicited in discussions with both the central bodies of the industry and with individual officers of the companies or works involved. In most cases, the information provided from the latter source has necessarily involved various shades of personal involvement and often a significant measure of post-rationalization. Some degree of cross-checking has been possible, but it cannot be asserted that in every or any case the crucial factors in the decision have been unequivocally delineated.

We begin by collecting in Table 22 the relevant information on the organization, location, production and potential future production of the principal steel makers in the United Kingdom. The figures cited as potential production in 1965 and 1970 are derived from the existing capacity and the expected capacity of schemes which have been approved by the Iron and Steel Board. There may be a certain increment to these figures, especially for 1970, arising from schemes which have been submitted but not yet been formally approved, and from those which may come forward in the next few years. It is expected that these may add at least a further 3·1 million tons.

In Table 23, the present situation concerning oxygen steelmaking processes in operation in the United Kingdom is summarized. In addition to those works noted, a number of other open hearth operators employ oxygen to a greater or less extent, but except possibly in the case of The Steel Company of Wales these activities can scarcely be described as intensive oxygen practice. The Guest Keen Iron and Steel Works at Cardiff is included as a typical case of intensive oxygen open hearth practice. Recently, Dorman Long (Steel) Limited has announced plans to install an L.D. plant with the added prospect of an F.O.S. unit as well. A Kaldo unit intended for the production of low phosphorus foundry iron is being installed by Stanton and Staveley Limited and is expected to be operational by 1965.

The nominal capacities shown in Table 23 are intended primarily to convey the general scale of the installations. Nominal capacity and operational capacity may differ considerably as the linings in L.D., L.D.–A.C. and Kaldo vessels are progressively consumed and, in the case of Ajax furnaces, it may be preferable in some cases deliberately to operate at a capacity below the nominal value in order conveniently to cope with slag foaming or other requirements. The figures quoted should not, therefore, be used to calculate accurate annual production levels in the several installations listed.

Comparison of Tables 22 and 23 reveals that there is no very obvious correlation between the level of production of crude steel by a company and its adoption of one or more of the oxygen steelmaking processes. Moreover, there is no clearly defined location pattern, though it is evident that works operating on high phosphorus pig irons derived primarily from home ores show some preference for Kaldo units. This is, however, only one of the several factors which has influenced the choice of a particular process. Another consideration is the time at which the decision to install a new steelmaking

Table 22: *Organization and production of steel in United Kingdom, 1960–70*[1]
(in thousands of tons)

Group	Company	Location	Production		Potential production		Products
			1960	1963/4	1965	1970	
Colvilles Limited	—	Glasgow, Motherwell, Glengarnock	2,159	2,190	3,430	3,430	Billets, plate, sections, rails, sheet, alloy steel products
Consett Iron Co. Limited	—	Consett, Co. Durham	992	911	1,102	1,323	Billets, plate, light rolled products
Dorman Long (Steel) Limited	—	Middlesbrough	2,327	1,852	2,565	2,565	Billets, plate, sections, rails, wire rods, strip
Duport Group	Briton Ferry Steel Co. Limited	Briton Ferry, Wales	250	270	300	300	Billets
	Llanelly Steel Co. Limited	Llanelly, Wales	245	193	240	240	Billets
English Steel Group	—	Sheffield	570	505	869	869	Billets, plate, bars and forgings, including alloy
Firth Brown Limited	—	Sheffield	182	142	185	200	Alloy sections and forgings
	Wm. Beardmore & Co. Limited	Glasgow	69	66	80	80	Billets and forgings
G.K.N. Steel Co. Limited	Brymbo Steel Works	Brymbo	197	288	325	350	Billets including alloy
	Guest Keen Iron & Steel Works	Cardiff	797	855	1,000	1,000	Billets, slabs and sections
	Lysaght's Scunthorpe Works	Scunthorpe, Lincolnshire	668	747	950	950	Billets, sheet bars, wire rods

Company	Works	Location					Products
Hadfields Limited	—	Sheffield	132	119	150	165	Alloy sections and forgings
John Summers & Sons Limited	—	Shotton	1,325	1,561	2,040	2,040	Sheet and plate
Lancashire Steel Manufacturing Co. Limited	—	Manchester	603	634	701	701	Billets, sections, wire rods
The Patent Shaft Steel Works Limited	—	Wolverhampton	245	314	335	384	Plate, sections
Richard Thomas & Baldwins Limited	Ebbw Vale Works	Ebbw Vale, Wales	803	929	1,075	1,075	Sheet, plate, tinplate
	Spencer Works	Newport	—	1,246	1,664	1,664	Sheet, plate
	Redbourn Works	Scunthorpe, Lincs	532	656	688	800	Billets, slabs
	Panteg & Gowerton Works	Swansea	242	201	235	235	Billets, sheet bars, light sections
	Other West Wales Works	West Wales	347	—	—	—	—
Shelton Iron & Steel Co. Limited	Subsidiary of John Summers & Sons Limited	Stoke-on-Trent	255	219	344	344	Heavy sections
Skinningrove Iron Co. Limited	—	Saltburn	277	257	290	290	Heavy sections, rails
South Durham Steel & Iron Co. Limited	—	West Hartlepool	1,080	1,280	2,000	2,000	Plate, heavy sections, rails
Steel Company of Wales Limited	—	Port Talbot	2,814	2,322	2,979	2,979	Sheet, plate, tinplate

Table 22: Organization and production of steel in United Kingdom, 1960–70[1]—continued
(in thousands of tons)

Group	Company	Location	Production		Potential production		Products
			1960	1963/4	1965	1970	
Stewarts & Lloyds Limited	Corby Works	Corby, Northants	1,145	1,123	1,375	1,560	Tubes, pipes
	Bilston Works	Wolverhampton	487	629	700	825	Tube steel
	Clydesdale Works	Glasgow	277	167	330	330	Tubes, pipes
Tube Investments Group	The Park Gate Iron & Steel Co. Limited	Sheffield	433	483	775	850	Bars, sections, strip
	Round Oak Steel Works Limited	Wolverhampton	452	495	505	528	Sections
The United Steel Companies Limited	Appleby-Frodingham Steel Co.	Scunthorpe, Lincs	1,374	1,496	1,704	1,819	Billets, plate, sections, bars, rods
	Steel Peech & Tozer	Sheffield	1,158	895	982	1,355	Billets, bars, sections, strip, tyres, wheels, axles and forgings
	Workington Iron & Steel Co.	Workington	325	304	302	340	Billets, rails, railway material
	Samuel Fox & Co. Limited	Sheffield	423	465	500	575	Alloy billets, bars, rods, sheet, strip
	Total of above		23,185	23,814	30,720	32,166	
	Other companies		1,120	936	1,025	1,239	
	Total all companies		24,305	24,750	31,745	33,405	

[1] Table 63 and Map Appendix II, *Development in the Iron and Steel Industry, Special Report, 1964.* Iron and Steel Board, H.M.S.O.

Table 23: Operational oxygen steelmaking plants in United Kingdom, 1964

Company or Group	Location	Type of Plant	Nominal Capacity tons
Colvilles Limited	Ravenscraig (Motherwell)	L.D.	2×110
Consett Iron Co. Limited	Consett, Co. Durham	L.D.	2×125
		Kaldo	2×120
G.K.N. Steel Co. Limited	Guest Keen Iron and Steel Works, Cardiff	Open hearth	Intensive use of oxygen
Richard Thomas & Baldwins Limited	Ebbw Vale Works, Ebbw Vale	L.D.–A.C.	2×45
	Spencer Works, Newport	L.D.	3×100
	Redbourn Works, Scunthorpe	Rotor	100
Shelton Iron & Steel Co. Limited	Stoke-on-Trent	Kaldo	2×45
Stewarts & Lloyds Limited	Corby	L.D.–A.C.	3×100
Park Gate Iron & Steel Co. Limited	Sheffield	Kaldo	2×75
Appleby-Frodingham Steel Co.	Scunthorpe	Ajax	5×300
G.K.N. Steel Co. Limited	Lysaght's, Scunthorpe	L.D.–A.C.	2×75

plant was taken. In the earlier years of these developments, it was widely held that the L.D. process was the appropriate one for low phosphorus pig irons and the Kaldo for high phosphorus bearing irons, these being the materials for which the respective processes were initially developed. A little later, with the advent of L.D.–A.C., the area of choice changed somewhat; and a little later still, as has been pointed out elsewhere, it was recognized that two slag practice with lump as distinct from powdered lime addition to the L.D. process would enable a rather wider range of pig irons to be treated than had originally been contemplated. Against this fluid technical situation, it is hazardous to evaluate too categorically in retrospect the quality of individual decisions without regard to time.

In the paragraphs which follow, we shall examine in more detail a number of the cases cited in Table 23, in order to elicit some of the varied factors influencing the decisions that were taken.

Colvilles Limited, Ravenscraig. This new works at Ravenscraig near Motherwell began production in 1957. The originally planned steelmaking facility comprised three basic open hearth furnaces of capacity in the range 250 to 300 tons, with provision for a fourth furnace to be added later. At the time of this original decision, the modern open hearth furnace was judged to be the best process then existing on the world scene. The possibility of installing the L.D.

process was, however, briefly considered, but was rejected then as being at too early a stage of development. No provision was made for the use of oxygen in the open hearth furnaces and, indeed, none has ever been employed to the present time. Initially, steel produced by the open hearth furnaces was intended primarily for use in the manufacture of plate. To this point, the development of the Ravenscraig works was essentially a conventional one co-ordinated with the company's other works in the same area.

The decision to introduce oxygen steelmaking was not one taken in isolation, nor was it associated in any integral way with the originally established facilities at Ravenscraig. It arose as a part of the so-called strip mill project which was instituted in 1958 as the result of a decision by the Macmillan government and involved the provision of a £50 million Treasury loan. This project, completed in 1962, comprised the construction of coke ovens, sinter plant, two additional blast furnaces, a fourth open hearth furnace, an L.D. steelmaking plant and a semi-continuous hot strip mill to roll finished strip up to sixty inches in width.

The study of the particular steelmaking process to be employed began in February 1959 with a visit of a technical team to Europe. The Rotor furnace was never seriously considered and the conditions appropriate for the Ajax process did not exist. The possibilities were therefore reduced to the L.D./L.D.–A.C. on the one hand and the Kaldo on the other. It was recognized from the beginning that, though one additional open hearth furnace could be added to the existing melting shop, this would not provide the necessary additional capacity and, furthermore, that the required extra capacity could be provided more cheaply in terms of capital by either of these oxygen processes. There was some pre-existing preference in capital term for the L.D.-type process.

Though the strip mill project was conceived separately from the existing plant, it was obvious that it would ultimately have to be related to and to some degree integrated with these facilities. In particular, this raised the issues of scrap availability, the phosphorus content of the pig iron to be employed and the possible alternative usage of oxygen converter and open hearth steels in both plate and strip manufacture. None of these considerations was wholly inflexible since, for example, though most of the scrap likely to be available in the area would be circulating works scrap, there were satisfactory prospects of bringing in imported scrap from elsewhere in the

United Kingdom or overseas at an acceptable cost in relation to that of pig iron. Moreover, the phosphorus content of the pig iron was to some extent related to the blast furnace practice employed and its cost, a lower phosphorus bearing material being obtainable at a somewhat higher price. The situation that has currently been achieved is that open hearth steel can be used alternatively in plate and strip without restriction, while the converter steel may to some extent but not universally be utilized in these two product areas. It has also been found technically and economically advantageous to utilize a somewhat lower phosphorus pig iron than had originally been intended, and to operate the open hearth furnaces with a 65/35 and L.D. vessels with a 75/25 hot metal to scrap charge.

The point that is emphasized by these considerations and events is the comparative flexibility of the Ravenscraig situation. In these circumstances, it would have been very difficult to make a case for the Kaldo in preference to the L.D./L.D.-A.C. process.

There were, however, some additional internal and external factors, not the least of which stemmed from the original decision on the strip mill project. In essence, the Macmillan decision provided for the construction of two strip mills—one at Ravenscraig and the other in Wales—at a time when there existed a case for only one such mill. It follows that, initially and probably for a considerable period, there would be excess national capacity leading to a low level of plant utilization. While in these circumstances it would be unrealistic to apply normal commercial criteria of judgment in the short term, this situation would strongly predispose those making detailed decisions on particular plant units towards minimum capital costs. There were also certain long-standing traditions in Colvilles which would tend to minimize operating costs. It is, for example, fairly widely recognized in the industry that in their open hearth practice this company had achieved very economical operating costs with minimum manning scales.

The organizational structure of Colvilles provides for comparatively few staff as distinct from line positions. The small technical team in whose hands was firmly placed the responsibility for recommending a particular steelmaking plant was led by one of the directors and comprised people who would be very much concerned with operating the plant once it had been built. In addition, the existence of three modern open hearth furnaces and the provision of the fourth, all of which were readily capable of producing steel suitable for strip making, removed the absolute necessity to depend wholly at the outset on converter steel, thereby affording the possibility of a period of some experimentation with the new converter process.

In the event, most of these considerations have applied to a greater or less extent. The strip mill in 1965 is still operating at rather less than half its design capacity, there have been some changes in operating techniques with the converter which is now employed wholly as an L.D. unit with a triple nozzle oxygen jet. The degree of interchangeability of open hearth and converter steel for plate and strip has been progressively defined and extended.

Consett Iron Co. Limited, Consett, County Durham. The situation here offers some contrasts with that at Ravenscraig. The Consett works had its origins in the small, individual blast furnaces in pre-steel days built to utilize a comparatively small pocket of local ore which has long since ceased to have any importance. The local coal, for so long an important raw material, is also now petering out, though there are adequate deposits in not too distant parts of the county. The conversion here from wrought iron to steelmaking did not pass through a Bessemer stage, but went directly to acid open hearth furnaces which were later replaced by basic units. For most of its life, this steelworks has been concerned with producing shipbuilding plate partly for local industry, and though today this continues as an important product, billets and small sections rolled at Jarrow are other significant outlets.

A special feature of the Consett situation is that it is a steelworks in comparative isolation supporting a town which has grown up over the past century. The siting of the works has little to commend it, as it is now not significantly dependent on local coal, not dependent at all on local ironstone, it does not enjoy the advantage of being on the coast or navigable river and in topography the site is rather unsuitable for a modern integrated steelworks. It is not too surprising, therefore, that in the 1930's there were very grave doubts if the works would survive the economic pressures of that period, especially as its main product was plate for shipbuilding which was also in serious difficulties at that time. That it did survive appears to have been heavily dependent on social and political pressures derived from the unacceptable prospect of creating a ghost town a little too remote in those days from possible centres of other employment.

In a very real sense, the Second World War and its immediately preceding period were the salvation of the Consett works, and though it has continued actively in production since that time, it is possible that it would be one of the early establishments to be closed if a major surgical rationalization of the British steel industry were at some future time to be undertaken. However, the social implications made plain in the thirties remain, though their impact is decreased by improved transport enhancing the prospect of the town's conversion to a dormitory area for the more favourable, neighbouring, industrial development centre.

The open hearth shop was constructed in about 1925 to house the then newly built 65 ton open hearth furnaces. These were replaced immediately after the war by furnaces of 150 tons capacity, but further expansion in the 1960's along this line was structurally impossible. The first approach employed to increase production was the installation of duplexing facilities in the form of two bottom blown Bessemer converters to be used as pre-refining units for the open hearth furnaces which were also somewhat improved by the use of a certain amount of oxygen. Though this cannot be construed as other than a perfectly reasonable decision in the postwar circumstances, it may also contain elements of uncertainty for the future. Within three years, it had become evident that a new steelmaking facility was required and from about 1954–5 attention began to be directed specifically to this end.

The customary visits abroad, especially to Austria and Sweden, were made and the real possibilities reduced to L.D., L.D.–A.C. and Kaldo furnaces. Initially, it was assumed that the pig iron which was to be processed would contain $0 \cdot 7$ to $0 \cdot 8$ per cent phosphorus, but again, as in the case of Colvilles, this requirement was not unalterable; currently the pig iron being employed contains about $0 \cdot 2$ per cent phosphorus. The decision that was ultimately taken was a rather odd one judged in retrospect, since it involved the installation of both L.D. and Kaldo processes. In this, there seems to have been a number of competitive considerations. At the time, the emphasis was on treating a rather high phosphorus pig iron, so that if an L.D.-type process were to be adopted, it was more likely to have been L.D.–A.C. than L.D., though there is some evidence that it was recognized that an L.D. unit with two slag practice with lump lime addition might have met the requirements. At that time too, the L.D.–A.C. process had not been outstandingly successful and there were, quite properly, some doubts about its capacity to fulfil the

Consett needs. These considerations taken together made something of a case for the Kaldo process, supported by the added prospect of incorporating a higher percentage of scrap in the charge.

At this stage it appears that personalities and similar less tangible considerations tended to assert themselves. Two points of view emerged, one a modern, severely cost-oriented, production viewpoint and the other, a rather more conservative approach of the traditional steelmaker long practised in open hearth operation. Quite apart from any particular Consett factors, the latter viewpoint in general tended to favour the Kaldo process. This was not, as might appear, simply prejudice or stultifying conservatism, but also embraced an important element of forward thinking in relation to ultimate product type and distribution. In the longer term, common shipbuilding plate seemed likely to constitute a progressively smaller outlet, and even if structural plate remained a principal product, there were clear signs that specifications were tightening and specific steels were likely to occupy an increasing part of the order book. The prospects of consistently meeting such a pattern of requirements were initially intrinsically greater with the slower working Kaldo than with the L.D.-type process, though as time has gone on and experience increased, this differentiation has declined significantly.

The ultimate decision to install both processes has most of the hall marks of a stalemate with a significant element of strong personalities and, perhaps, emotional commitment. In retrospect, this decision may well be judged to have been wasteful in capital, but undoubtedly the issues were not as clear at the time as subsequently they may have appeared.

Guest Keen Iron and Steel Works, Cardiff. This works, one of the three in the G.K.N. Steel Co. group, dates essentially from 1935, though its origins go back to the 1890's. The complete rebuilding of the works in 1935 was based not unnaturally on open hearth practice; at the outbreak of the Second World War steelmaking facilities comprised two 100 ton fixed furnaces, three tilting units each of 200 tons capacity and two tilting furnaces each of 300 tons. This works, together with that of Stewarts and Lloyds at Corby built at about the same time, were the two modern works in pre-war Britain on which the early official attempts to calculate operating costs and profit margins for the industry were based.

The installation of the tilting furnaces at Cardiff—rather unusual at that time—arose because of the use originally of high phosphorus iron. Today, the ore employed in iron making comprises 20 per cent Oxfordshire ironstone with 80 per cent imported ore. Initially, the works was intended as a producer of ordinary low carbon steel, but during the war there was a growing trend towards the making of special steels. This has since continued, and today some 200 different specifications are regularly made. Equipped with both fixed and tilting furnaces of modern design, this works was well equipped to meet this change. The big range in composition (0·06–0·85 per cent carbon) and the wide variety of applications arising from the use of 90 per cent of the steel produced within the G.K.N. group are important factors which point to the steady evolution in steelmaking techniques rather than the adoption of quite new converter processes. This is a policy that has been consistently followed since the end of the war. In view of the good but rather restricted site, substantial capital investments in materials handling facilities have been made and it is evident that the choice of priorities in capital investment has been a rational one.

The open hearth furnaces were largely rebuilt in 1947 and converted from gas to oil firing. Experiments involving the introduction of oxygen began in the early 1950's and in the period, 1956–58, lancing with oxygen was consistently used; oxygen was not, however, employed in the combustion air. This pattern was closely similar to that followed at the Steel Company of Wales with which there was free interchange of information. Hand-held, consumable mild-steel lances inserted through furnace doors or other openings present some hazards to the melters and, at all events, are largely a short-term expedient. In 1958–9, experiments with retractable roof lances and with variations in lance and nozzle design were actively pursued and have continued in an evolutionary way ever since. As the oxygen rates were progressively stepped up, the roof lining lives decreased; the acid roofs were replaced with basic suspended roofs and a further increase in oxygen usage introduced; the roofs were raised until they virtually touched the cranes and oxygen usage again stepped up. At the same time oxygen-oil burners were installed in the roofs to decrease melt down times.

The significant point here is that there has been a quite deliberate policy of step wise innovation, each problem being solved or alleviated in turn until the limits imposed by materials or geometry have been reached. In this, the Guest Keen Iron and Steel Works have been following in many respects an outlook strongly held and successfully implemented by the Steel Company of Canada to the extent that, by adopting a fuelless practice during oxygen refining in a very large open hearth furnace, this company has essentially been operating a very large L.D. vessel in the shape of an open hearth. For this purpose, the Steel Company of Canada used a specially constructed furnace, but apart from this the approach at Cardiff has been of a similar kind.

Given the situation which existed at the Cardiff works in respect of location, area, product pattern and outlets and the physical and capital nature of the existing facilities, there is little doubt that this approach has been an appropriate one. Ultimately, there must come a stage at which it becomes physically, technically or economically unprofitable to proceed further in this direction and at this point important new decisions will have to be made.

Richard Thomas & Baldwins Limited. The three works in this group involved are those at Ebbw Vale (Wales), Spencer (near Newport) and Redbourn (Scunthorpe), where three different choices, respectively, L.D.–A.C., L.D. and Rotor have been made. Though it might have been expected that there would be substantial liaison between these works of a nationalized company, this does not seem to have occurred to any substantial extent. The Spencer works is probably the most straightforward of the three cases. This was conceived as a new, large scale works on a green fields site to be operational in 1962. The main product was to be strip, and though the initial capacity was set at $1 \cdot 4$ million tons, there was from the outset a firm intention to expand as conditions demanded. The situation was, therefore, an unusual one in the British steel industry and differed somewhat from the Colvilles' development at Ravenscraig.

Although experiments at Ebbw Vale in 1960, in which one of the existing Bessemer converters was replaced by an L.D. vessel with top blowing with oxygen, were designed ostensibly to gain operational experience in this process for the new Spencer works, it is difficult to track down the extent to which transfer of experience occurred. Indeed, it appears that the character of the proposed works at Spencer and the fact that this was to be a steelworks in the new image more or less automatically directed attention to the L.D. process. Other processes such as the Kaldo were scarcely seriously considered. The Spencer works was intended to utilize pig iron made from imported ore and likely to be low in phosphorus. This factor also contributed to the choice of the L.D. steelmaking process.

The installation has not, however, been without its operational problems. The use of the originally supplied simple nozzle led to considerable skulling around the vessel neck, but this has now largely been overcome by substituting a triple nozzle lance originating from Japanese practice. Here and elsewhere, there is evidence that the dissemination of information through the L.D. licensors has been less than entirely satisfactory, and that semi-independent Japanese developments of the process often appear to have been adopted to overcome operational problems. One particular feature of the Spencer units is that they operate on very hot metal, and as a result are able to absorb unusually large quantities of scrap (up to some 30 per cent).

The Ebbw Vale works is one which has owed something to or been afflicted by political considerations and can scarcely be described as being well located. Its pre-war steelmaking facility comprised three Bessemer converters and one open hearth furnace originally with the object of employing a duplexing operational technique. Initially, the intention was to treat high phosphorus iron. In the late 1950's, a major renovation scheme was undertaken and this included, among other things, the conversion of the three existing Bessemer converters to oxygen/steam blowing, the installation of a fourth converter and the rebuilding of the three, then existing open hearth furnaces, each being increased in capacity from 90 to 120 tons. By July 1960, one of the existing Bessemer converters was replaced by an L.D. vessel which was the first L.D. unit to be commissioned in the United Kingdom. By 1964, the whole of the Bessemer plant had been replaced by L.D.–A.C. converters, the provision of lime dispensing being included to afford flexibility in respect of the phosphorus content of the pig iron to be refined.

In this case, the several decision steps are readily delineated. The Bessemer experience extended to oxygen-steam blowing, then to top blowing with oxygen and ultimately to total replacement by L.D.–A.C. units traverses the full pattern of steps on a single site. The final provision of L.D.–A.C. rather than L.D. units probably represents little more than a simple precaution in line with the practice which was increasingly preferred elsewhere following the successful introduction of powder addition on a production scale. Clearly, where there is any significant prospect of lime addition being required, the increased capital cost of providing this facility at the outset is likely to be minimal.

The decision to instal a 100 ton Rotor furnace at Redbourn appears

14

somewhat out of keeping with the trend of opinion on the world steel scene. It will be recalled that the Rotor furnace was initially intended as a pre-refining rather than as a steelmaking unit. We have already seen in several of the previous cases discussed that there was a more or less distinct period or stage of evolutionary development in which with an existing works the production capacity could conveniently be increased somewhat by the addition of some form of duplexing or pre-refining. There is a limit to the extent to which increased production can be achieved in this way at an acceptable level of additional capital expenditure, but given a substantial element of doubt concerning the permanence of increased demand, such an approach is in no sense unreasonable. To a quite significant degree, the economic conditions in the late 1950's and early 1960's, as reflected in the steel production figures noted in an earlier section, afford some justification for this course of action. There are, of course, other methods of pre-refining or duplexing which might have been rather simpler, cheaper and, if the increased demand proved transitory, more readily reversible. The question of why the Rotor furnace was chosen therefore remains largely unanswered by this line of enquiry.

A second approach of some validity in this case concerns the individual or group of people making the initial decision. In circumstances where there exists a development group or section rather independent of the production line management, any recommendation arising therefrom may be subject to critical scrutiny by those concerned with production, so that at board level somewhat conflicting views may have to be resolved. Alternatively, if there is no such staff group and the recommendation flows from a production oriented line group, a board, especially one remote from the operational centre, may be confronted with a rather different situation. It is often under these circumstances that a forceful advocate may carry the day more by force of personality than by rational argument.

A further possibility is that there was a positive decision to invest money in an experiment and, here again, such a decision will in many instances be based as much on judgment of the sponsor himself as on the detail of the proposal. Whatever the detailed situation at the particular time at which this decision was taken, and this can no longer be accurately reconstructed, it appears that a strong personal element was operative.

Even to the casual visitor there are some distinct differences in the organization and character of Richard Thomas and Baldwins in comparison with other steel companies or groups of nominally similar size. In the privately owned companies, it is not uncommon to find a director leading the technical team charged with making a proposal on a new steel plant. There is in these circumstances small prospect of the system itself exercising any significant effect on the decision, though, of course, the existence of a director of dominant personality may well result in a largely personal view or assessment being adopted. In the case of R.T.B., there exists a number of additional staff groups, nominally research and/or development sections, whose operating responsibilities are none too clear. For example, there is at Pye Corner a substantial group whose main function is process research and development associated with the Spencer Works, but it is uncertain if this group would be able to exercise any decisive influence on, say, technical decisions relating to the next stage of expansion of these works. This is not offered as a criticism of the particular group cited, but rather to underline the contrast between the very direct decision making line in, for example, Colvilles and the apparently more complex and to the outside observer rather amorphous structure in Richard Thomas and Baldwins. The critical point is that any potential advantages that may accrue from operating with a development team which is substantially independent of the production line management are likely to be largely vitiated in the absence of an effective communication route.

Shelton Iron and Steel Co. Limited. This old established works at Stoke-on-Trent is one of the smallest of the larger companies in terms of production and, unlike the larger group of John Summers and Sons Limited with which it is affiliated, has traditionally been a maker of heavy sections. In 1959, the introduction of new standard sections made it imperative that the old rolling mill be replaced. The iron making facility on the works was still satisfactorily operational, but the open hearth shop was near the end of its useful life. Of the three basic units, two were therefore due for immediate or near-immediate replacement and in these circumstances a good case could have been made for closing the works altogether. It operated on Northamptonshire ore yielding a pig iron containing $1\cdot4$ per cent phosphorus and there was adequate coal in the area, but beyond this it was a works in comparative isolation operating in circumstances where substantial expansion was not likely.

With any long established works, whatever the particular social environment, there is always some reluctance to close it down. In the

Stoke area this argument should have had comparatively little force, since in no sense would the modest increase in unemployment resulting from such a decision have created unresolvable hardship at a time when other adequate employment opportunities were available. The alternative to closure was an imaginative reconstruction across the whole area. The decision that was taken was of this kind and called for the installation simultaneously of an oxygen steelmaking plant, a continuous casting unit and a new rolling mill to be operated in conjunction with the existing ironmaking facility. Such a decision was apparently the only real possibility in terms of an acceptable return on the capital expenditure and appears to have resulted, at least to some extent, from the fact that its chief protagonist had begun his steel career at Shelton and had retained an affection for the works. There is, even in many of the toughest of steel men, a sentimental streak and though this may seem a relatively trivial point, it is one likely to arise when the alternative was closure of the works.

Because of the high phosphorus content of the available iron, the real possibilities in regard to steelmaking were at that time the L.D.–A.C. and Kaldo processes. The decision was ultimately made in favour of the latter, but not without a comparative, detailed study of both. Three points seem to have been paramount. The first was that in trials conducted in Luxembourg with the L.D.–A.C. process, the prospects of being able to operate consistently by catch carbon techniques did not appear sufficiently encouraging. The second was that because of the decision to install a continuous casting unit, it was necessary for its successful operation to achieve close temperature control at the end of the steelmaking stage. This was judged to be more likely with the Kaldo than with the L.D.–A.C. process. A third consideration was additional capital cost involved in providing a waste heat boiler with an L.D.–A.C. plant, a requirement which was unnecessary if a Kaldo plant was installed. An additional factor was the prospect of being able to utilize more scrap in the Kaldo than in the L.D.–A.C. process.

Given the decision to continue production at Shelton involving the provision of these three new facilities, the choice of the Kaldo at that time in preference to other converter process appears justified. If, however, the decision had been made a few years later, it may have gone the other way, though it must be conceded that with continuous casting in its comparative infancy in Britain, requirements arising from this more than those from any other source seem likely to have sustained the original choice.

The decisions at Shelton should also properly be considered in relation to the parent company, John Summers and Sons Limited. The scale of production by the latter at Shotton is in the vicinity of 1·5 million tons per annum which goes largely into sheet for the automotive industry. The proposed scale of the new developments at Shelton, on the other hand, amounted to 350,000 tons per annum, which is small by national and international standards and without attractive prospects for rapid growth. One element in the decision to re-equip the Shelton works appears to have been that this would afford experimental experience within the group in both oxygen steelmaking and, of greater significance, continuous casting. This element was apparently a significant one in securing the Iron and Steel Board's approval, for without it the rebuilding of a small works of this kind runs somewhat counter to the general policy of rationalization.

Stewarts and Lloyds Limited, Corby. The Corby plant was built in 1935 adjacent to the Northamptonshire iron ore field specifically to make Bessemer steel for tube making skelp and in this respect stands somewhat apart from other steelworks in Britain. Moreover, because of the long period of Bessemer practice, new developments were likely to start from a somewhat different background from that operating in many other centres in Britain. The first stage comprised the installation of a tonnage oxygen plant—the first to be owned and operated by an iron and steel producer in Britain—and the modification of the Bessemer plant for oxygen enriched and oxygen/steam blowing. At the same time, an extensive programme of modernization covering many phases of this works was undertaken. By 1961 as part of this programme, the decision was taken to expand the productive capacity at Corby substantially by building a new steel plant.

The intention at that time was that this be an L.D.–A.C. plant and towards this end a single converter unit was installed for full scale experimental work. Following this programme, a firm decision had been taken by 1964 to replace the existing Bessemer plant by an L.D.–A.C. facility comprising three 110 ton vessels.

There are two points of special interest in this case. In the first place, the long experience of Bessemer converters would have strongly predisposed the decision towards an L.D.-type plant in preference to a Kaldo installation which would have greatly increased the cycle time over that to which the works had become accustomed in Bessemer days. The second point was the quite deliberate experimental programme on a full scale within the Corby works. We have noted other cases in which the experimental work preceding decisions consisted largely of *ad hoc* trials in Sweden on the Kaldo, in Austria on the L.D. and in Luxembourg on L.D.–A.C. Of these, there appears to have been some dissatisfaction with the arrangements, timing and conduct of trials on L.D. and L.D.–A.C. on behalf of potential customers from the United Kingdom, whereas the impression created by the Kaldo demonstrators appears to have been universally favourable both in technical and social terms. One suspects that at least in some situations where a potential licensee has been debating between Kaldo and L.D.–A.C., this factor has operated in the favour of the former. The decision of Stewarts and Lloyds at Corby to undertake their own trials locally with an experimental converter represents not only an admirable degree of Scots canniness, but is also consistent with a predisposition towards the faster L.D.-type process.

Park Gate Iron and Steel Co. Limited, Rotherham. In the latter part of the 1950's, a decision was taken to build on a site adjacent to the existing works at Rotherham a new integrated steelworks with a design capacity of 365,000 tons per year. In the first stage, iron was to be drawn from the existing blast furnaces and the new installation involving a blooming mill, continuous billet mill and narrow strip mill would roll the total production of the two works of some 800,000 tons per year. The steelmaking facilities in the old works comprised ten basic open hearth furnaces, five with capacities in the range 60–79 tons and five of 80–99 tons. The pig iron available was high in phosphorus ($1 \cdot 1$–$1 \cdot 3$ per cent) and the ultimate products in bars, sections and strip covered a wide variety of specifications from dead soft to the high carbon types. A further general requirement was the need to absorb a large proportion of scrap.

The choice here lay essentially between L.D.–A.C. and Kaldo and the final decision rested primarily on the results of trials conducted in Luxembourg and Sweden. The experience of the technical team involved in these trials was that while the L.D.–A.C. could in a satisfactory proportion of cases meet the required specifications for phosphorus, the performance with respect to carbon content was unsatisfactory. In these early experiments in 1956–7, some trials were also carried out with the L.D. units at Linz, but, as might have been expected, the phosphorus requirements could not be met. On the other hand, a series of twenty-four heats aiming at carbon contents between 0·08 and 0·8 per cent carried out on Kaldo units in Sweden were successful in all respects. The decision in favour of the Kaldo process was therefore a straightforward one based on these trials.

In retrospect, one may well question whether this simple go-no-go criterion involving limited trials on someone else's plant is a wholly adequate one. There is little doubt that if this approach is adopted, the slower working Kaldo unit by its very nature is more likely in isolated experiments to be able to achieve the desired ends, whereas considerable detailed experience and practice may be required before the same results can be achieved with acceptable consistency with L.D.-type units. This is not to suggest that in the circumstances the decision at Park Gate in favour of the Kaldo was wrong or ill advised, but rather to emphasize the possibility of false criteria emerging when the difference between long-term potential and short-term actual performance is not able to be adequately assessed within the limited scope of the available experimental study.

Appleby-Frodingham Steel Co., Scunthorpe. In many respects the situation here is rather similar to that which existed at the Guest Keen Iron & Steel Works at Cardiff. There existed large modern tilting open hearth furnaces designed to treat high phosphorus pig iron; at the same time, the necessary technical ability was available supported by considerable enthusiasm and confidence in local innovatory efforts. In 1955–6, the joint problems of lancing with oxygen, the issue of roof life and the alternatives of silica and basic roofs were explored. From these and similar studies, there ultimately emerged the modified tilting open hearth furnace adapted for intensive oxygen usage and prescribed methods of operation which together constitute the Ajax process. The difference in approach between the Guest Keen works at Cardiff and Appleby-Frodingham lies mainly in matters of local detail; in the former, the step-wise process was a steady, continuous one without major, recognizable, close-ended stages, while in the latter, the earlier experimental work in co-ordinated form emerged in a distinct jump as a circumscribed process justifying a separate name and capable of sale or transfer to other sites. It must, however, be noted that no other operators in the United Kingdom or elsewhere have taken up the Ajax process as such, though it is available for exploitation, presumably because of the engineering problems associated with conversion of furnaces which may be arranged somewhat differently from those at Scunthorpe.

The new capital cost involved in the conversions at Appleby-Frodingham is stated to be £180,000 per furnace with an average time of twenty-eight days. The first Ajax furnace came into production in January 1961, and since then four other furnaces have been converted or are in process of being converted. This largely speaks for itself. One of the distinctive elements in these events at Scunthorpe has been the personal contribution of Mr A. Jackson. That enthusiasm and technical competence are essential ingredients in such an operation is obvious, but, in addition, Mr Jackson has highly developed skills in advocacy and communication as his two books[1] and many papers and lectures abundantly demonstrate.

G.K.N. Steel Co. Limited, Lysaght's Scunthorpe Works. This long established works dating from 1910–12 has since that time steadily increased its steel output. Pig iron produced from a mixture of local and Northamptonshire ore and running about 1 per cent phosphorus has for more than fifty years been refined in open hearth furnaces of both fixed and tilting types. In the 1950's, open hearth practice was intensified in the customary ways by the use of oxygen both in combustion and to lance the bath to accelerate carbon removal, by the introduction of higher fuel input rates, and by the substitution of basic refractories in the furnace roof to permit the use of higher furnace temperatures. With these operational changes to the existing capital equipment, steel production was increased to an average of nearly 700,000 tons per year.

To add at least a further 300,000 tons of productive capacity beyond this enhanced open hearth figure, a new steelmaking plant was required. The initial choice was based on a study of available processes begun in 1958. Because of the high phosphorus content, a two slag converter process was necessary, the possibilities including L.D.–A.C., Kaldo or Rotor. The choice of L.D.–A.C. in preference to either of the other two appears to have been based primarily on a desire to keep the engineering situation as simple as possible and a considerable degree of confidence that the L.D.–A.C. process could be made to fulfil the local needs. There were no trials in Europe with any of these several processes, though a series of visits to various installations was, of course, made. This confidence has been fully

---

[1] *Steelmaking for Steelmakers*, The United Steel Companies Ltd., Sheffield, 1960; *Oxygen Steelmaking for Steelmakers*, G. Newnes Ltd., London, 1964.

justified and in the early commissioning trials it was established that the new plant could make the whole range of Lysaght's steels hitherto made in open hearth furnaces.

In designing the plant, some interesting degrees of freedom have been incorporated. In Stage I, two blowing stations were built, but with only one of the two vessels to be in operation at a given time. In addition, provision was made for removing a vessel to a relining station, so that when in Stage II a third vessel was available, the initial rated capacity of 450,000 tons per annum could be virtually doubled. On completion of Stage II, therefore, the proposed total works production target of 900,000–1,000,000 tons per annum could be met by the L.D.–A.C. plant alone, and the open hearth plant could then be dispensed with.

A feature here is the way in which there is an inbuilt provision of time, if this were required, to achieve the degree of operational experience necessary to make consistently all the required grades of steel on the new plant. At the same time, a programme of data handling and ultimately process control by computer has been embarked upon at the outset, thereby ensuring that the plant in addition to producing acceptable steel also generates a maximum amount of information about itself and its operational peculiarities. The decisions here are remarkably clear cut and incorporate in this leap-frog principle an important element of insurance. The only temporary disadvantage lies in the need to provide a second blowing station rather earlier than immediate requirements demand. This appears to be a small cost to bear in the light of the other accruing advantages.

SUMMARY

We have now outlined the circumstances relating to the introduction of eleven oxygen steelmaking plants in the United Kingdom. Of these, two are concerned with the intensive use of oxygen in adapted open hearth furnaces, and there are three L.D., three L.D.–A.C., three Kaldo and one Rotor installation. The last named is comparatively unusual both in Britain and elsewhere; the choice of L.D. where the iron to be treated is low in phosphorus is an obvious one needing little further amplification. The dual choice of L.D. and Kaldo at Consett, as has been pointed out, appears to have been the result, among other things, of strongly pressed but divergent views of competing groups. The more or less even split between

L.D.–A.C. and Kaldo in cases in which a high phosphorus iron was to be treated reveals some interesting sidelights. In situations where self confidence has been high or built-in insurance could be effected, for example at Corby and at Lysaght's Scunthorpe Works, the decision has been for the rather more uncertain L.D.–A.C. process, but where these circumstances were absent or heavily modified in other ways, as at Shelton and Park Gate, the decision has been in favour of the Kaldo. The Kaldo units have suffered from significant mechanical troubles and their higher capital cost and lower output suggest that with growing operational experience on L.D., modified L.D. and L.D.–A.C. units, these will in the future account for a steadily growing proportion of steels of many kinds made in the United Kingdom.

Some conclusions

We now bring together some of the salient points which have emerged and draw some conclusions. In so doing, it is necessary to reiterate two qualifications, the first that steel is not a single material but a generic term covering a multitude of different types, and the second, that the iron and steel industry encompasses an enormous range of activities stretching, for example, from mining at one end to the production of tin plate at the other. In these circumstances, wholly valid generalizations are likely to be so general as to be value-less, while those of some usefulness or significance are almost certain to be seriously inapplicable over more than one sector of the industry. The comments which follow are more specifically directed to the area of steelmaking.

In terms of natural resources, Britain today is not in a specially favoured position as a steelmaker. While coal resources are adequate in quality and quantity and are spread fairly widely over the country, many of the seams are rather thin for modern mechanical mining methods in comparison with the situation in some other countries. Local deposits of iron ore are limited in quantity, not widely dispersed and generally lean in quality. Imported ore has for nearly a century been an important commodity, but this need not, as a recent experience in Japan has abundantly shown, be an insuperable problem. The well established, rationalized purchasing arrangements under B.I.S.C. (Ore) have largely ensured continuity of supply, control of quality variation and stabilization of prices for this imported ingredient.

Steelmaking in Britain has had a long and rather checkered history. In the nineteenth century, this country was the source of a significant part of the world's supply of steel and initially the home of steelmaking methods epitomized in the Bessemer converter and the open hearth furnace, both of which were British inventions. Since these methods came into common use in the 1860's and 1870's, the pattern of innovation in steelmaking has been one of evolution deeply rooted in a craft tradition. The First World War did little to alter or influence this pattern and it was with the hallmarks and values of this tradition that the industry faced the changed economic conditions of the 1920's.

The British iron and steel industry as it then existed failed to match effectively the open competition of this period, though not wholly for technological reasons, and from this point in time onwards the industry has been largely in a defensive position. It is true that this period produced some degree of internal reorganization, combination and rationalization, but the industry was sheltered by heavy tariffs before this activity had reached an advanced stage. In some respects, the British chemical industry faced similar problems at this time, but here the response in terms of rationalization went much further with the emergence in 1926 of Imperial Chemical Industries Limited, which has continued to the present day to compete effectively in many of the world's markets. The reason for this is not, however, to be explained simply in this way.

By the time the wartime controls were finally removed in 1953, the British iron and steel industry had been operating for nearly a generation without the stimulus of effective, external competition. For this and other reasons, the remnants of the pre-1914 entrepreneurial spirit seem to have been largely and perhaps irrevocably dispersed.

The defence of the craft tradition in steelmaking has been furnished by many hands and many factors, by managements, the trades unions, the technical nature of the processes, the government, the universities and, indeed, British society at large. The change from owner-managers to professional managers brought comparatively little change, perhaps because of the latter only an insignificant proportion had had experience outside the industry. The trades unions industrially and politically and against a background of deeply rooted fears of unemployment, short time, outright redundancy and a substantial body of ancient grievances have, until quite recently, strongly defended their fortress against innovation. The technical nature of the steelmaking process, tolerably well understood in general terms, but still relatively unexplored in detail, happens to be of a kind which does not easily lend itself to study in isolation in research institutes and universities. In any case, the outlook in British universities has been directed strongly away from seeking for its own sake an understanding of the basic processes of traditional industries, and there have been no institutes corresponding in character, size or ethos to the Technische Hochschule at Aachen or Carnegie Institute at Pittsburgh. The policies of successful British governments have emphasized the social responsibilities of the iron and steel industry, often sustaining an existing works or an industry pattern in the face of a changed total situation or, as in the case of the Macmillan government, of sponsoring two strip mills where, in commercial and economic terms, one would have sufficed. British society in the large has, at least since the Second World War, been permissive in mood rather than critical in outlook. There remains much devotion to a glorious but rather faded past and a considerable lag in the widespread recognition of Britain's changed role in world affairs—culturally, scientifically, economically and politically. One of the most striking pieces of evidence for this lies in the circulation figures of such newspapers as *The Daily Express.*

It is tempting to suppose that the various systems of supervision by the state to which the British iron and steel industry has been subjected since the early activities of the I.D.A.C. in the 1930's have been a major factor in shaping the character of the industry and its approach to innovation. In regard to capital expenditure, it remains substantially true that companies have had to raise their own capital for expansion, replacement or modernization (though there have been cases of government loans or other forms of assistance) and, except for the short period of public ownership, there is little evidence that other forms of supervision have seriously hindered or altered the extent or the pattern of investment. This is in part a speculative judgment, but in support of it there is the inescapable fact that the same people or the same sort of people have been involved in its detailed direction and management irrespective of the form of government supervision. Again, it might be argued that the setting of controlled, maximum domestic prices in the war years and subsequently under the aegis of the Iron and Steel Board has been a factor of significance in shaping the financial and commercial decisions in the steelmaking industry. But this scarcely seems to have been the case in any decisive sense. The methods employed in calculating these prices, in principle at least, offer no discouragement to the erection of modern, efficient plant, but by the same token they have not influenced in a very marked way the closing down of obsolescent units, especially where these have long since ceased to have significant capital values and where interest and depreciation components in production costs have been minimized. Moreover, even if there had been much less government supervision of the industry in the past thirty years, many of the co-ordinating functions carried out by such bodies as the Iron and Steel Board are likely to have been exercised by the British Iron and Steel Federation itself. The early methods of arranging price increases and the *ad hoc* committees to co-ordinate capital expenditure in the pre-war period can scarcely have developed into a system greatly different from that operated by the Iron and Steel Board. Whether such arrangements under the B.I.S.F. would have been in the public interest—whatever detailed connotation is placed on this phrase—is not a particularly relevant question, but to the question of whether it would have been in the interests of the industry, the answer is unequivocally in the affirmative.

The catastrophe of the 1920's and early 1930's left its mark indelibly on the industry in Britain. From that time forward, there steadily emerged the policy of trying more and more closely to match home demand with home production. Since that time, too, many other countries which were traditional importers of steel have set up their own steelmaking facilities, also with the avowed intent of meeting all or nearly all their domestic needs at least in common types of steel. In brief, these types of steel in crude and semi-finished form are decreasingly a deliberately made export commodity. There will, of course, always be some world trade in this area because at any point in time there will be some countries which have not achieved internal sufficiency and, even if this state were broadly reached, to the extent to which home demand and production are never universally and continuously quite balanced. The point that we here emphasize is that the general policy of estimating as closely as possible the expected home demand, of matching this by productive capacity leaving only a little surplus capacity to cope with unscheduled increments and, if necessary, being prepared to sell abroad a small surplus possibly at marginal rates is one which is dictated by the world scene in steel rather than by any decision of the British or, indeed, any other government. In 1963, the world export trade volume in steel was 10 per cent of the total volume of production and 8 per cent of the then existing capacity. By 1970, the corresponding figures are expected to be 8·8 and 7·6 per cent.

Such a policy carries with it in practice a number of important consequences. The first is the need for considerable skill in predicting steel demand in a situation in which many of the steel using industries will be affected both by domestic conditions and by those in the export markets for an immense variety of products. Secondly, the domestic cost of making steel in Britain necessarily has to be fairly closely in line with that abroad, especially in Europe, a condition which implies that there be no very serious lag in Britain in introducing cost-reducing technical innovations and practices. This does not, of course, mean that the domestic price of British made steel can or should ever be made to match that of steel imported at marginal prices in times of world surplus. Thirdly, that a stop-go economic policy in which the rate of industrial expansion in times of boom is extraordinarily high will necessarily pose additional, substantial, and possibly insoluble production problems not only for the steel industry, but also in the supply of other basic commodities such as chemicals. And fourthly, there will be continuing need for a high degree of co-ordination in the common steelmaking sector of the industry to maintain and sustain the required levels of productive capacity and appropriate distribution of product types.

In Table 20 the listed prices for British made steel were compared with those for a number of other countries. These figures refer to 1963 and, at that time at least, a reasonable parity was being achieved. It also is of some interest to extract from Tables 2 and 13 the tonnages of steel made by the L.D. process relative to the total steel production. The results of this calculation are shown in Table 24. While these

Table 24: Proportion of total steel made by L.D. process

Country	Steel production 1963 (short tons) (Table 2)	Installed L.D. Capacity 1964 (short tons) (Table 13)	% L.D. Capacity/ Steel production
U.S.A.	109·1	16·6	15·2
E.C.S.C.	80·6	15·8	19·6
Japan	34·7	18·9	54·5
U.K.	25·2	4·4	17·5

figures refer to specific years, take no account of the rate of installation of L.D. plants and include no reference to other oxygen steelmaking techniques, they serve to illustrate that, in respect of the leading oxygen steelmaking process, the United Kingdom at the time in question was matching reasonably well the performance of both U.S. and E.C.S.C. Japan is a rather special case resulting from the delayed start in re-establishing its steel industry.

The task of the control and reduction of production costs in an established industry can, however, be achieved in many ways in addition to the introduction of new processes. This is likely to be particularly important in steelmaking where the initial capital cost of plant is high, its life long and its value periodically enhanced by relining and refurbishing, and where the materials cost is a significant component in the overall total. Areas of materials handling, fuel economy, organization and methods study, process control and operational research, to mention a few, all offer prospects of fruitful economies. There is no doubt that the utilization of oxygen in open hearth furnaces and the consequential tightening up of the whole steelmaking operation in these furnaces has had a significant effect on reducing the costs of steelmaking generally in Britain and improving the output rate in relation to the capital investment.

There has been some public criticism of the British steel industry on the grounds that it should have developed its own oxygen steelmaking processes rather than utilizing the L.D., Kaldo and similar processes initially developed abroad. It is one of the peculiarities of some sectors of British industry and opinion generally that to import technological know-how is in some way thought to be sinful or disloyal. The circumstances in which the L.D. process was developed in Austria were unusual and may not have been reproduced elsewhere. They certainly did not exist in the United States or the United Kingdom in the postwar years. The position in Sweden where the Kaldo process emerged is of particular interest. This country had certain distinctive natural resources from which its industrialization proceeded. One of these was the extensive deposits of high quality iron ore on which the Swedish iron and steel industry was based and from which ultimately stemmed a variety of special quality tool, ball bearing and stainless steels and products often of distinctive design made from these materials. Sweden has a small population—some seven millions—and, like the Dutch and the Swiss, the Swedes have deliberately concentrated their industrial innovation and their research and development effort in specific fields pertinent to their circumstances. The widely known brand names of Alfa Laval, Facit, Electrolux, Volvo and A.S.E.A. serve to illustrate the success that has attended their efforts. It is against this wider pattern that the Kaldo developments should properly be considered.

In Britain, the resources of qualified scientists and engineers are such that to imagine that this country can depend exclusively on its own inventions and innovations is both invalid and dangerously misleading. One of the tasks that has to be seriously faced is the selection of areas in which the limited research and development component of this resource can most profitably be deployed. It is not the purpose here to argue the proportions of this effort which might be utilized in various technological areas or industries, but rather to emphasize that within the total scope of the iron and steel industry, there will inevitably be insufficient resources to man a local innovatory effort over the whole spectrum. For the purpose of this exercise, it is useful to divide this total span in two, conveniently referred to as bulk steelmaking on the one hand and special steelmaking and utilization on the other. The question that we now pose is that if the effort has to be concentrated predominantly in one of these two areas, which one is to be selected?

The facts of economic life in Britain, whether in the short or long term, are the need to achieve an adequate (and preferably steady) rate of growth without incurring a semi-permanent balance of payments problem. Let us, for the purposes of argument, first elect to place the major effort in the bulk steelmaking area. Such an effort might aim at eliminating the separate blast furnace and steelmaking furnace and replacing them by a single direct reduction process. This is a proposal which at the present state of knowledge is quite feasible in scientific terms, but would require a very substantial development effort to be technologically viable on a production scale.

We do not know in quantitative terms how the successful solution of all the problems related to a direct reduction process would affect the production costs of common steel. B.I.S.R.A. ran a research programme on direct reduction for some years, but has now discontinued it largely because the prospects of implementing a successful new process of this kind did not appear to be economically attractive. As we have seen in the case of oxygen steelmaking technology, the introduction of new but not wholly radical processes has almost exclusively occurred when an existing facility has literally worn out, or where additional productive capacity is required. In the latter situation, we have also noted the sustained efforts in a number of cases to modify existing plant and to evolve processes making use of new technology with a minimum of new capital expenditure. The proposition that existing, viable furnaces be replaced by a single stage, direct reduction process is unlikely to be economically feasible, unless this change catastrophically reduced the production cost of steel. The literal costs of processing through the blast and steelmaking furnaces are, however, only two of many components that make up the total cost of producing steel.

In all these circumstances, it is at present difficult to conceive that a fully developed direct reduction or other revolutionary process would enable the basic price of steel to be reduced by more than a comparatively small amount. Let us suppose, for the sake of argument, that a major research and development effort on new steelmaking processes was successful and, when implemented, resulted in a decrease of 10 per cent in the average price of £40 per ton. The effect of this reduction on, say, the production cost of a motor car using about a ton of steel would be negligible.

The alternative of substantially increasing exports of semi-finished common steels by a reduction in basic price of this magnitude is not a real possibility under present or foreseeable conditions. For the most part, steel sold on the export market is disposed of at prices well below domestic levels. For example, steel sheets currently sold in Britain at £54 per ton could be bought on the open world market in December 1965 for £37 per ton; one British steelmaker is reported in the press as saying that this price would not cover the direct costs of production. The prospects of a new steelmaking technology bridging this kind of gap and maintaining this situation for an extended period is, to say the least, remote. The choice of the first alternative does not look at all appropriate for Britain.

A similar quantitative effort in the area of special steel alloys with distinctive properties, of improved methods of using these materials and in the wider area of steel utilization, especially where the innovatory component including design is high and where there is a substantial added value in manufacture, is likely to produce far more intensive and widespread benefits within the industry, and make a more effective contribution to national economic growth and to the balance of payments problem. Furthermore, a research and development effort so directed would build upon an already established area

15

of achievement in science in Britain. Nobody pretends seriously that the situation discussed above is quite as concrete as has been suggested for the purpose of argument, or that any area of technical and productive endeavour can be so neatly cut in two, except perhaps in a totalitarian society. Nor does the proposition of concentrating research and development effort in selected segments imply a total neglect of the remainder.

In British steelmaking in the past decade, the pattern of development that has taken place is fully in keeping with the arguments advanced in the preceding paragraphs. Today, there exists a mixture of imported technology in the form of L.D., L.D.–A.C., Kaldo and Rotor plants, and the results of local evolution of the intensive oxygen usage in existing open hearth furnaces without major capital expenditure. Within the total national and international framework of steelmaking and the economic circumstances and problems in Great Britain, this pattern is scarcely open to fundamental criticism, though points of detail may well be debatable.

While much has been done in recent years in improving the productivity of existing processes and in progressive, adaptive development of plants employing the newer steelmaking technology, continued gains in this direction rest, among other things, on identifying potentially useful areas of endeavour. One of these is illustrated in Table 25 in which are set out the steel conversion factors for non-alloy steels. Taken as a whole, these figures reveal that about one quarter of all ingot steel made is returned for reprocessing as circulating scrap. In comparison with other process industries, this must be regarded as very high. Continuous casting offers prospects of reducing this figure and is one of its principal advantages, but, apart from this possible solution involving a substantial capital investment, there does not appear to be wide recognition of this as an area where significant direct gains might be made by other methods. Perhaps this is an oversimplification, or is it an example of a piece of deep rooted steelmaking lore?

The ability to identify potential areas of process improvement and to effect appropriate changes depend, to some extent, on the nature of the recruitment into the industry. Compared with the science-based industries, the British iron and steel industry has for a long time pursued a rather restrictive personnel policy. This is not peculiar to the British scene, but is identifiable in varying degrees in iron and steelmaking in several parts of the world. The common core

Table 25: Conversion factors, non-alloy steels

Product	Ingot to Product	Product	Ingot to Product
Heavy rails, sleepers, fish plates, etc.	0·726	Cold rolled strip	0·684
Plates 3 mm thick and over	0·708	Sheets under 3 mm thick	0·748
Heavy sections and bars	0·793	Tinplate including blackplate	0·719
Colliery arches and light rails	0·733	Tubes, pipes and fittings	0·688
Wire rods	0·772	Tyres, wheels and axles	0·551
Other light sections and bars	0·757	Forgings excluding drop forgings	0·553
Hot rolled strip	0·755	Steel castings	0·498
Bright bars	0·717	Other forms	0·759

Weighted average 0·736

of this policy is a strong tradition that the man who has not poured hot metal, whatever may be his technical or personal qualities, is to some degree an outsider. This tends to produce a situation in which the bulk of the recruitment of technical staff in the production area is at a very young age. In-service training and experience are highly valued, while external technical or scientific education acquired before or during the early years of employment is often regarded as more in the nature of a useful but slightly superfluous decoration. One of the effects of this outlook is that while there is some movement of professional personnel from one company or works to another, there is very little interchange of staff with other industries. It might, for example, prove very interesting to have staff at the plant superintendent level exchanged between some areas in the chemical industry and the steel industry, or, alternatively, to observe the effects of using professionally trained chemical engineers instead of metallurgists on steelmaking plants.

One of the important though less publicised contributions which B.I.S.R.A. had made in this area has been to function as a recruiting route for graduates from a wide variety of disciplines. A number of these cases was identified and, from this small sample at least, it appears that graduates entering the production side of the industry in this way apparently do so successfully and achieve a satisfactory and sometimes outstanding rate of progress thereafter. Against this heartening development must be set the recently advertised proposal of the Appleby-Frodingham Steel Co. to recruit sixteen year olds from the local grammar schools into a sandwich course covering or substituting for the last two years of a normal secondary school course. If the object of this scheme is to improve the quality of the increasingly important technician group, it has something to commend it, but as a means of training technologists who will be useful critics of established practice, it leaves much to be desired.

The contribution which B.I.S.R.A. has made and can make to the improvement of existing processes is a major one, especially where specific studies or trials on the plants of member companies are undertaken. The difficulties of smoothly organizing and carrying through to a successful conclusion this kind of process research are not insignificant and in practice such projects are seldom quite as straightforward as it might seem on paper or in committee. Such efforts have the added value of introducing a spirit of enquiry where perhaps none existed before and in a subtle way of catalysing other efforts initiated wholly within the particular works or company.

The development of tonnage oxygen has been described at some length,[1] although this is peripheral rather than central to the main topic. Part of the value of its inclusion lies in the contrast it provides between the pattern of development of oxygen steelmaking and that of tonnage oxygen itself. While the acceleration in the commercial production of tonnage oxygen undoubtedly has resulted from its rapidly growing usage by the iron and steel industry, there is little evidence that the latter acted as a stimulant in its development. It is not surprising that the chemical industry has played a larger initial role, since industrial gases are generally classed as part of that industry. The scientific base was here clearly recognizable and largely complete, and the development problems to be solved were identifiable against formally defined theoretical limits. There are, in fact, few examples in which the interplay of theory and practical ways and means are so well demonstrated. Part of this situation stems from the fact that the process involved is a physical separation contrasting markedly with the complexity of the chemical reactions of steelmaking.

Tonnage oxygen also provides an almost classical example of the economies of scale. There are a number of particular reasons for this, notably the zero or near zero materials cost of the input, the identifiable variable costs of production, the continuity of the operation under steady state conditions, the absence of materials handling problems and the neat circumscription of the total process. These are circumstances in which accurate calculations can readily be made on the economies that will result from operation on a large scale and, in addition, of the rate of variation in ultimate product cost with scale.

[1] See pp. 115, 123 above.

This kind of situation is fairly common in the chemical industry, especially in plants involving liquid and gaseous reactants and products.

One of the several arguments advanced to support the proposal to nationalize the basic steelmaking segment of the British iron and steel industry is that it will become necessary to increase vastly the size of production units in order to achieve significant cost reductions. It is perfectly true that, in general, the capital cost of process units is not a linear function of their capacity, for example, a 20 h.p. motor costs less than twice that of a 10 h.p. one. There have not, however, to the author's knowledge been any reasonably complete, published calculations of the direct application of these principles to the particular area and problems of iron and steelmaking. The reasons for this are not difficult to find and among them we may note the following:

(a) the number of situations in which a green fields development of a fully integrated works has been undertaken is small and the prospects of adequately testing scale equations under truly comparable conditions have been limited;

(b) the successive stages of burden preparation, iron making, steelmaking, casting, soaking, rolling, shaping and finishing, either individually or collectively, can scarcely be described as continuous processes in the sense that an oil refinery, an oxygen plant or an ammonia synthesis unit may be said to operate continuously;

(c) the relationships between scale and capital costs for items ranging, say, from a sinter strand to a rod mill and involving materials handling facilities such as cranes, materials assembly, buildings, soaking pits, ingot trains, oxygen plant, blast furnace, steelmaking facility, to mention a few, are likely to be quite different at a given co-ordinated level and, further, to display a wide range of sensitivities with scale itself;

(d) in situations in which the materials component in the final product cost is substantial, where the product mix is varied widely, and where the technology is tolerably stable, the gains in cost in relation to scale may prove not to be as clear cut or as substantial as might be hoped by simple extrapolation from, for example, the chemical process industries.

Where detailed calculations for an integrated steelworks have been attempted, the graph of unit cost against scale so obtained is not a smoothly falling curve which obeys a simple power law. Such calculations do, however, show that in the United Kingdom considerable gains should accrue up to a scale of 1 million tons per year, somewhat smaller ones from 1 to 2 million tons, but beyond this latter figure the advantages become very uncertain. Though the economies of scale for an integrated steelworks have great political and philosophical attractions, these may not be fully borne out in practice. Despite this, the U.S.S.R. is said to be exploring the possibilities of integrated steel plants in the range from 12 to 24 million tons per annum.

V

SOME COMPARISONS

Although it is inherently unlikely that generalizations of wide applicability can be drawn from only three studies of the kind described in the earlier chapters, a detailed comparison should indicate similarities and differences which may be usefully borne in mind in further studies of a similar type. One approach would be to divide these case histories into four parts, prehistory, invention, development and exploitation, but such a concrete division may obscure some features; for this reason, we shall first try to look at these examples in broad outline and allow the several phases to emerge more naturally. One advantage of this approach is that it may provide a better method of relating oxygen steelmaking to the new industrial polymers with which, at least superficially, it may appear to have little in common.

One way to do this is to discount initially the identity of the particular new products and processes which are central to their respective studies and to set out each narrative in a linear, chronological series with appropriate notations of all the events which have been recorded. If this is done, it immediately becomes evident that the series on steelmaking falls into two quite distinct cycles separated by a considerable interval of years. The first of these extends from the period before the early 1850's, when the principal ferrous materials of commerce and industry were wrought iron and cast (pig) iron, through to the establishment of the four, main steelmaking methods, acid and basic Bessemer and open hearth processes, which were well established by the end of the nineteenth century. In the fifty years from about 1880 to 1930, apart from the introduction of electric steelmaking in a limited way for common low carbon steels, there ensued a period of stable technological practice which may be taken as the background for the second cycle extending to the present time. With this subdivision of steelmaking, there are in all four series now to be considered.

In order to compare systematically these four series, corresponding features of similarity between them are first sought by point by point analysis. The first of these to emerge is that at which it can be said a new idea crystallized from or was injected into the then existing situation. In the two steelmaking series, Bessemer's invention of 1855 and the experiments on oxygen top blowing begun by Durrer in

Berlin in 1933 are two such indices. It is not vital whether we associate the date of 1930 or 1933 with the latter, but, as Durrer carried out his first experiment in 1933, this is preferred. In the Terylene series, there are two possibilities, one in 1922 when Staudinger began his classical work on macromolecules and the other in 1928 marking the commencement of Carothers' programme at du Pont. The first of these is preferred, because Carothers acknowledged the importance of Staudinger's work, especially on polyoxymethylenes, as pointing the way to the rational use of particular types of monomers in order that the synthesis of the polymer would yield a product of unequivocal constitution. For polythene, the corresponding time is 1919, this being the date of Freeth's report and the occasion of his first visit to Leiden and the year in which Michels began in Amsterdam his work on the isotherms of gases at high pressures. There do not appear to be any serious competitors for this choice.

Noticeably, the Bessemer datum is for a specific invention, whereas in the other three series these indices are effectively the starting points of research programmes. The justification for this superposition lies in the recognition that in the middle of the nineteenth century invention in a craft industry is as near as one can get to the initiation of a new idea within a framework of existing knowledge and practice. The other three are in good correspondence, though the character of Durrer's work was largely technological, while the others were more in the nature of programmes of laboratory scientific research. However, Michels' work involved the solution of serious technical as well as detached scientific problems. It is also widely recognized that Michels was a far better experimental than theoretical physicist.

We next seek in the four sequences the events which marked the ends of these invention or research stages at which it can be said with some certainty that development as a distinct entity could begin, or at which the central invention was made. Those selected are: the Thomas invention of 1878, the decision at VÖEST in 1949 that they had the necessary information to embark on the building of a viable oxygen steelmaking plant, the Whinfield and Dickson invention of 1941 and the synthesis of polythene in 1935. The first three of these are not open to serious challenge, but for the polythene sequence it might be argued that the 1933 synthesis by Gibson and Fawcett would be a better choice. This argument is rejected because it is quite clear from the record that only after 1935 was it recognized that a possibly significant invention had been made.

In the intervals between these sets of two fixed points, there exist in three of the series a single distinctive intermediate point. In the first steel sequence, this is the invention of the open hearth furnace by the Siemens and the Martins in 1869; for Terylene, it is the invention of Nylon in 1938, though the first laboratory synthesis should probably be placed a little earlier than this; for polythene, there is the Gibson and Fawcett synthesis of 1933. For the second steel series, there is no formally similar, distinctive event, but it is apparent that sometime between 1945 and 1949 it became widely recognized that tonnage oxygen for steelmaking was a real possibility in economic terms. Though it is somewhat unreasonable to give to this point a specific date, it is convenient for our present purposes to do so by choosing 1947 as the mean between the extremes of the period. With this inclusion, all four series are reduced to a similar interval pattern.

These twelve points of correspondence, three in each of the four series, have been used to construct Fig. 3 in which they are labelled in chronological order B, D and F. The resulting intervals are described by A for the prehistory, C and E for the successive parts of the research and invention phase, and G for the undiscriminated region of development and exploitation. In G, no attempt is made to show fine chronological detail, but only to illustrate in broad terms the emergence of the several avenues of exploitation.

This representation focuses attention on the lengths of the corresponding time intervals. Phases A and G are open ended, but for the closed periods, C and E, the intervals in order are, respectively, 14,14,16,14, and 9,2,3,2 years. With one exception in the second group, these figures are surprisingly consistent and prompt a more detailed comparison of the character of the events which have been set in like positions. There is, for example, a good deal in common between Thomas' invention of basic steelmaking and Whinfield and Dickson's discovery of Terylene. The former was based upon a clear recognition in scientific terms of the problem of refining high phosphorus pig irons. The successful solution was a direct result of this assessment and was made by a man who by his own efforts external to the iron and steel industry had acquired the necessary chemical knowledge. Whinfield also formulated his problem in clear chemical terms and directly from this was able to propose a means by which a possible fibre-forming polyester might be obtained using symmetrical aromatic instead of aliphatic groups in the polymer chain. This work, like that of Thomas, was conceived largely externally

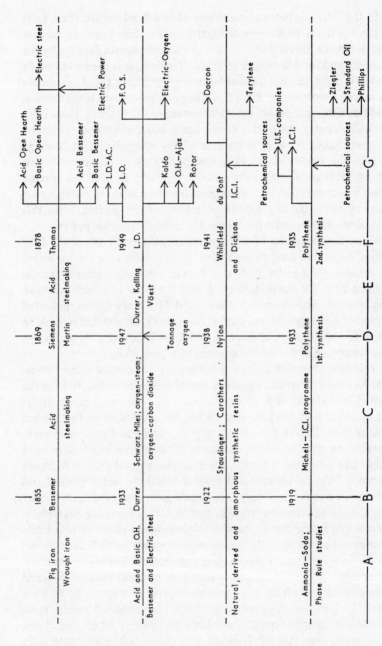

FIG. 3 Comparison of case studies on matched time scales

from the inventor's day to day interests in The Calico Printers' Association. Moreover, while Thomas' invention of basic linings was initially intended for use in Bessemer converters, it was rapidly applied also to the Siemens-Martin furnace; in the case of Terylene, du Pont, already the inventors of polyamide fibre-forming polymers, was quick to seize upon Whinfield's polyester as soon as they were informed of its existence. In Fig. 3, the Nylon invention is properly shown as having the same relation to the Whinfield and Dickson discovery as the Siemens-Martin furnace bears to Thomas' invention.

A broadly similar situation applies in the case of the L.D. process, but with somewhat less distinctiveness. The relevant facts are in three parts: (i) the patents of Schwarz and Miles in 1939 for top blowing of molten pig iron with oxygen as a means of making steel; (ii) the successful experiments of Durrer and Hellbrügge in March 1948, which established that steel in ton quantities could be made in this way; and (iii) the successful trials on an experimental scale carried out by the VÖEST group at Linz in 1949. Here the steps are smaller and the progress more evolutionary, but it is clear from the record that Suess and colleagues at Linz appreciated the importance of the positioning of the jet, the design of the nozzle and its geometry and the selection of the oxygen rate. There was a deliberate assessment of the variables which were likely to be important and an experimental assessment of their effects in a way not different in principle from Whinfield's extension of polyesters from symmetrical aliphatic to symmetrical aromatic systems.

There is also some similarity between the Linz work and the second polythene synthesis. Neither was wholly novel; Suess and his group at Linz and Perrin and his team at Winnington approached their respective tasks knowing that they were not breaking completely new ground; both teams had improved facilities over those that existed a year or two previously; and in both cases, there was considerable emphasis on getting the technique right. It is tempting to make too much of the type relationships of the terminal points F; it should, however, be noted that the identification of them is largely a retrospective judgment by the external world after subsequent industrial exploitation had taken place.

The events designated at point D have some likeness, though this is less striking than the terminal points F. The parallel between the Siemens Martin–Nylon and Thomas–Whinfield situations has already been mentioned and is not pressed further. It might have been expected from their broadly similar character that polythene and Nylon emerged in the first instance under like circumstances, but a wholly convincing case for this is difficult to make. The first synthesis of polythene was essentially an accidental observation, to which not much significance was attributed at the time, made in the course of a programme of research on the effect of high pressures on chemical reactions. In contrast, Carothers' intentions, stated in his letter to Johnson written in 1928, unambiguously included not only the possibility, but something of a prediction that the condensation of a diamine and a dicarboxylic acid of particular form would probably result in a linear polymer of high molecular weight. One such combination became the synthetic route to Nylon 66. But it has to be remembered that Carothers turned to polyamides only after he and his colleagues had done a considerable amount of work on polyesters from aliphatic acids and these had not yielded industrially interesting materials.

This comparison prompts a closer examination of the Wilmington and Winnington research teams. In the former, there was no doubt who was the leader—Carothers without question; at Winnington there was no equivalent character of the same stature. To some extent, Michels constituted a nucleus in the early stages, especially as regards technique, but he was essentially an external consultant and contributed only to one aspect of the programme. Freeth, Rintoul and Cocksedge had administrative influence and responsibility, but none was engaged in the detailed work; Swallow was also to some extent on the periphery, though his contribution as an anchor man is easily underestimated; Gibson and Fawcett and Perrin and Williams and possibly Paton were on much the same level. In the event, it is not possible to judge one team as having been more successful than the other, for both produced a polymer of signal, long term, commercial importance.

These considerations lead to speculation on whether polyethylene terephthalate might have been a du Pont invention had Carothers not died in 1937. The group at Wilmington was built around Carothers and derived much of its impetus from him, so that his death created a vacuum which was almost impossible to fill. At Winnington, no such problem arose and the succession of Gibson by Perrin was not only feasible, but was carried out with little apparent fuss. The change appears to have given some enhanced stimulus to the programme, though this was not its primary object.

The programmes in phase C are characterized by differences rather than similarities. Of the period of steelmaking between 1855 and 1869, we know comparatively little in detail, except that the Bessemer process was taken up vigorously in the United States where the pig iron was suitable and that there was a strong demand for steel rails for the then rapidly expanding railway system. In Britain, the emphasis was on investment in suitable iron ores in Spain and their transportation to the United Kingdom. Steelmaking in this period and in the years which followed was very much an art, the size of individual plants was small, communications were not highly developed within the industry with the result that the pattern was predominantly a local evolutionary one. Much more is known of the period, 1933–47, in respect of oxygen steelmaking. Here, the use of oxygen enriched blasts, oxygen-steam and oxygen-carbon dioxide mixtures were also largely evolutionary steps. The Schwarz and Miles patents seem to have fallen largely on deaf ears, probably because very few steelmakers before the war would have seriously contemplated the question of whether oxygen would be available in the near future in sufficient quantity at an acceptable cost. While the war stimulated the advance of gas separation technology and enlarged its potential scale, steelmaking in wartime Europe generally meant trying to maintain production from existing plants. In the United States, the aim was to increase production in existing plants and, where necessary, to construct quickly additional plants to a proven design. In these circumstances, it is not surprising that there were no major excursions into revolutionary steelmaking methods.

These two C phases in steelmaking are likely to differ from the other two in nature, scale and location. Even today, research on a laboratory scale into potential steelmaking processes is apt to contribute less to the creation or stimulation of new processes than is the case within a large sector of the chemical industry. The steelmaking sector of the iron and steel industry is concerned with the production from a limited range of natural resources of a basic material on which manifold variations in composition can be imposed. In contrast, the chemical industry deals with a large number of wholly distinctive products often of unique chemical and structural constitution also made from a limited number of raw materials, but by a large miscellany of methods. Effective research in this area can thus be carried out against a background containing many degrees of freedom. Research in and for the chemical industry tends to fall into two phases:

(i) laboratory work directed to the synthesis of new chemical compounds of potential interest and the discovery of new synthetic routes to compounds of proven interest; and (ii) the conversion of selected cases into methods suitable for industrial use. The first of these activities is to a large extent missing from the iron and steel industry with the result that semi-technical, pilot- and full-scale studies tend to predominate; because of their nature, they tend to merge imperceptibly with the development phase. This is not, of course, to say that there are not needs or adequate opportunities in this field for the application of specialist scientific or technological knowledge, for example, chemical thermodynamics, heat and mass transfer, diffusion theory and so on.

Generally speaking, research on a laboratory scale is much cheaper than semi-technical and pilot-plant investigations. For this among other reasons, the chemical industry is able to undertake the former with some facility without initially worrying too much about the problems which might arise in the conversion of possible new processes discovered in this way to an industrial scale. At this laboratory level, it is feasible to indulge in speculative work in a way which is seldom possible on a larger scale. The Carothers' work and the research programme which ultimately led to polythene are fairly typical illustrations. That the process which emerged in the latter case required almost a new kind of chemical technology at pressures an order of magnitude higher than those previously employed had no inhibitory effect on the research programme itself. These kinds of issues underline some of the differences which exist between the potential of research expenditure to induce technological change in these two industries. They also point to the cognate desire for a rather higher return on capital investment in the chemical compared with the steel industry and to the emphasis on evolutionary process changes which tends to characterize the latter.

One area of the chemical industry, especially in the United Kingdom, which has reached a maturity comparable with steelmaking is the segment concerned with the ammonia-soda process primarily used for the production of soda ash (sodium carbonate). This is a monopoly of Imperial Chemical Industries and has reached a stage of maturity at which any research and development effort is properly now concerned almost entirely with the problems which arise from the design and operation of larger and larger plants. Typically, the return on capital in this sector is distinctly lower than in many other phases of the chemical industry, the prospects of competition are minimal and the monopoly has not been a subject of significant public criticism. Indeed, the situation has now been reached where it is pertinent to ask how long I.C.I. will go on meeting the British demand for soda ash at a profit margin below par? Any answer to such a question has, however, to be related to other aspects of the alkali industry centred in the joint production of electrolytic caustic soda and chlorine and to the multitude of other products which have made rapidly increasing demands for chlorine in the past decade.

A further point of interest is that, at the times the Carothers' work and the I.C.I. Alkali Division's programme on high pressure chemistry was instituted, both sponsors were operating profitably and could comfortably make resources available. The commonly quoted argument, that firms should embark on such research programmes when business is slack, or when their financial position and future are in jeopardy, has a nice theoretical ring, but seldom works out quite that way in practice. It is, however, worth noting that The Calico Printers' Association maintained their research programme throughout their long period of financial stringency, though it cannot be claimed that Whinfield's excursion into polyesters formed any central part of this effort.

The phases designated by E in Fig. 3 were in some measure determined by what had occurred up to point D. In all four series, this period involved a change of personnel, either wholly or partly. Thomas, Kalling and the VÖEST group, Whinfield and Dickson and Perrin, Williams and Paton were essentially newcomers to their respective activities. In three of the four cases, polythene being the exception, there was also a change in environment, but the nature and extent of these changes differ somewhat in degree. Thomas' thinking to all intents and purposes was wholly external to the steel-making environment; Durrer, Kalling and the VÖEST group were within the relevant industry a little removed from the centre; Whinfield and Dickson worked on the periphery of the chemical industry; and in the polythene case, there was a barely discernible break at the first synthesis in 1933. While these examples point to some advantages which seem to have accrued when fresh minds were brought to bear on the respective problems at particular stages, it would be incautious to derive general conclusions from them.

We now turn to the other end of the four series in Fig. 3 in which no detail of the development phases are shown. These are hard to distinguish in the two steel series because of the nature of the research effort which gave rise to the new processes and because there were comparatively minor questions arising from the nature of the products and the market demand for them. There were, of course, some uncertainties on whether L.D. steel could be used successfully for all applications for which open hearth steel had hitherto been preferred, though this was a less serious query with the Kaldo product. The product quality, especially as regards consistency, depends to some degree on the particular practice followed and the acceptance or otherwise of a product reflects, among other things, the conservatism of the consumer. However, the change from one kind of steelmaking to another necessarily takes place over a considerable period of time, except in unusual circumstances where much of the existing plant is destroyed by war, and there is adequate time and opportunity for adaptation and the development of the skills needed to meet detailed requirements. If a free market exists and if a basic material made by a new process is cheaper, the consumer has considerable incentive to adapt his processes to receive this cheaper feedstock. This situation has not, however, operated in the United Kingdom. With a changing pattern of home supply, there has been a tendency on the part of some consumers to favour imported material on the grounds of uniformity of quality, but this is likely to be of temporary significance.

In contrast, polythene stands at the other end of the spectrum, and while there existed commercially in the 1930's other synthetic resins, such as polystyrene, polyvinylchloride, polymethylmethacrylate, polythene possessed some distinctive properties which initially suggested its application for more specialist purposes. Its development was promoted but also complicated by the Second World War, and the particular circumstances which then prevailed may not be reproduced. Ultimately, three issues of importance emerged: (i) finding uses and hence markets; (ii) developing the necessary production and application technology; and (iii) securing cheap supplies of ethylene once polythene had emerged from the realm of specialist use. Only (ii) and (iii) have parallels in oxygen steelmaking and then to a far smaller extent.

Terylene tends to stand between these two extremes, for though there were here complications due to the war this, itself, heightened interest in synthetic fibres. Unlike polythene, Terylene scarcely had to create its own market from scratch, but only to find its effective place in the general scheme of natural, derived and synthetic fibres. The technology was not particularly unusual and from a quite early stage the implication was that large scale productive units would be required. Perhaps more than in the other cases, terephthalic acid from petrochemical sources was of greater significance than the need to secure supplies of oxygen or ethylene. In all three cases, however, development depended vitally on the availability of a major resource externally to the main line. This points to the related proposition that there are often advantages in initiating innovation by using special resources or a combination of them that may be available naturally or can be secured in other ways as a spring board for invasion of other technologies. There is little doubt, for example, that, although oxygen needs for steelmaking did not themselves generate the research and development effort in the 1920's and 1930's towards tonnage oxygen, the recognition of this possibility by the industrial gas operators has been a potent factor in developing in the direction of securing economies of scale. Similarly, the rapid growth in the demand for ethylene for polythene manufacture catalysed efforts to find routes to chlorinated hydrocarbons from ethylene instead of from acetylene. This in turn stimulated refinements in cracking and purification techniques aimed, in part, at producing both ethylene and acetylene balanced and adjustable with respect to the relative demands for these two streams. Moreover, in the United Kingdom the naphtha by-product of the petroleum refining industry became the important source material for these petrochemical processes.

One of the features of phase G is the diverse and sometimes competitive processes which multiplied from the original single stream. The L.D., L.D.–A.C., Kaldo, Ajax, Rotor processes and intensively oxygen-assisted open hearth furnaces were in large measure competitive, though initially they were aimed at treating particular sorts of pig iron or of continuing certain existing steelmaking practices. They all emerged over a comparatively brief space of time and, with the exception of the Rotor and Ajax processes, have been widely adopted. There are on the world steel scene few barriers to this spread of technology and the licence and royalty fees are generally quite low. The more or less competitive development of polyethylene terephthalate by I.C.I. and du Pont which created foci in Europe and the United States, respectively, from which other manufacturers drew their know-how, illustrates a relatively ordered pattern of the spread of innovation. The polythene case was complicated by the war and by the anti-trust judgments and pressures, though the ultimate result was not too different from that for Terylene. The Ziegler and, to a smaller extent, the inventions of Standard Oil of Indiana and Phillips Petroleum, though emerging as competitors, were not initially conceived in this way. The speed with which the Ziegler patents were taken up in Germany and the United States reflected the extensive and rapidly growing demand for polythene which had been built up in the preceding years. The early response by I.C.I. in developing a modified, high pressure process to make polythene of higher density more in keeping with the competitors and partly to offset the inroads by the Ziegler material was hardly surprising. It cannot be validly argued in this case that the limited monopoly conferred by the original patent inhibited other inventions and developments.

There may be something of a parallel to be drawn between the F.O.S. process for steelmaking and the Ziegler, Standard Oil and Phillips inventions, in that they sought to overcome a feature of their respective forebears which was in some circumstances inconvenient. For the moment, however, it is uncertain whether the F.O.S. will be another Ziegler, or will find only a limited number of exploitation venues.

Finally, it is tempting to conclude from the examples described in this book, that industrial innovation proceeds in a series of waves— a period of stable technology followed by an inventive burst which leads to subsequent development and exploitation with or without multiple or competitive branching—to be again followed by a period of stable technology until the next inventive burst occurs. One strong argument against such a simplified, general view is the diverse and interlocking character and externally stimulating effects of a burst of innovation in one area on other fields and forms of scientific and technological endeavour. More and more, it becomes difficult to separate categorically in terms of a standard classification one industry or one field from another. The beginnings of this are evident in the examples that have been discussed and this trend seems likely to continue with important implications for the economic, social and political organization of man.

APPENDIX I

Patents bearing the name of K. Ziegler issued before his appointment as Director of the Max Planck Institute for Coal Research at Mülheim.

Patent No.	Brief Title	Date	Remarks
Ger 512,882	Organic compounds of lithium	15.10.1929	H. Colonius, inventor
Fr 728,241	Alkylation of nitriles	14.12.1931	
Ger 570,594	Alkylating acid nitriles	15.3.1933	H. Ohlinger, inventor
Br 394,084	Acetamides as hypnotic agents	22.6.1933	
Ger 581,728	Alkylating acid nitriles	2.8.1933	Addition to Ger 570,594
Ger 583,561	Tertiary acid nitriles	6.9.1933	
Ger 601,047	Alkali amides	7.8.1934	
Ger 591,269	Cyclic cyanoketimides and ketones	19.1.1934	Schering Kahlbaum A.G., H. Ohlinger and H. Eberle, co-inventors
Fr 774,316	Alkali metal amides	5.12.1934	
Ger 611,374	Substitution products of acetamide	27.3.1935	cf. Br 394,084, co-inventors H. Eberle and H. Lüttringhaus
Ger 615,468	Substituted alkali metal amides	6.7.1935	
Ger 620,904	Cyclic cyanoketimides and ketones from alkylene dinitriles	30.10.1935	Addition to Ger 591,269, Schering Kahlbaum A.G.
Ger 624,377	Derivatives of cyclic keto carboxylic acids, ketones	23.1.1936	Addition to Ger 591,269, Schering Kahlbaum A.G.
U.S. 2,049,582	Acid amidines	4.8.1936	Rohm & Haas Co.
U.S. 2,068,586	Organic cyclic cyano compounds	19.1.1937	Schering Kahlbaum A.G.
U.S. 2,068,284	Cyclic keto carboxylic acids	19.1.1937	Schering Kahlbaum A.G.
U.S. 2,103,286	Compounds of β-keto-carboxylic acids	28.12.1937	Schering Kahlbaum A.G. cf. Ger 624,377
U.S. 2,141,058	Substituted alkali metal amides	20.12.1939	Schering A.G.
Ger 671,840	Cyclic aryl ethers	14.1.1939	K. Lüttringhaus, co-inventor

SELECT BIBLIOGRAPHY

Polythene and Terylene

BOOKS

ANON., *Fifty Years of Calico Printing*, The Calico Printers' Association Limited, Manchester, 1949.

ANON., *Landmarks of the Plastics Industry*, Imperial Chemical Industries Limited, London, 1962.

BOLITHO, H., *Alfred Mond: First Lord Melchett*, Secker, London, 1933.

COHEN, J.M., *The Life of Ludwig Mond*, Methuen, London, 1956.

CROSS, C.F. & BEVAN, E.J., *Researches on Cellulose*, vol. I, 1895–1900 (1901); vol. II, 1900–1905 (1906); vol. III, 1905–1910 (1912); vol. IV, 1910–1921 (1922), Longmans Green, London.

CROSS, C.F., BEVAN, E.J. & BEADLE, C., *Cellulose*, Longmans Green, London, 1895.

DONNAN, F.G., *Ludwig Mond, F.R.S.: 1839–1909*, Institute of Chemistry of Great Britain and Ireland, London, 1939.

ENOS, J.L., in *The Rate and Direction of Inventive Activity*, Princeton University Press, Princeton, N.J., 1962.

GIBSON, R.O., *The Discovery of Polythene*, Royal Institute of Chemistry Lecture Series, no. 1, 1964.

HARDIE, D.W.F., *A History of the Chemical Industry in Widnes*, I.C.I. Limited, London, 1950.

JEWKES, J., SAWERS, D. & STILLERMAN, R., *The Sources of Invention*, Macmillan, London, 1957.

KAUFMAN, M., *The First Century of Plastics*, The Plastics Institute, 1963.

KRESSER, T.O., *Polyethylene*, Reinhold, New York, 1957.

MARK, H. & WHITBY, G.S. (eds.), *Collected Papers of Wallace Hume Carothers on High Polymeric Substances*, Interscience, New York, 1940.

MITCHELL, J.W., *Chemistry and Industry*, Jubilee Lecture to the Society of Chemical Industry, 908–35, 29 May 1965.

MUELLER, W.F., in *The Rate and Direction of Inventive Activity*, Princeton University Press, Princeton, N.J., 1962.

OLD CALICO PRINTER (*pseud.*), *Some Thoughts on the C.P.A.—A plan for adventure and a more sporting spirit*, George Falkner & Sons, Limited, Manchester, 1929.

RENFREW, A. & MORGAN, P. (eds.), *Polythene*, Iliffe & Sons, London, 1957.

ROYAL SOCIETY, *Obituary Notices of Fellows: C.F. Cross*, **1**, 1932–5, p. 459.

—— *Obituary Notices of Fellows: Sir William Jackson Pope*, **3**, 1939–41, pp. 291–324.

SUITS, C.G. & WAY, H.E. (eds.), *The Collected Works of Irving Langmuir*, vol. 12, Pergamon, London, 1962.

WATTS, J.I., *The First Fifty Years of Brunner, Mond and Co., 1873–1923*, Brunner, Mond Co., 1923.

JOURNALS

ADAMS, ROGER, *National Academy of Sciences of the United States of America Biographical Memoirs*, **20**, 293–309, 1939.

ANON., *Rubber and Plastics Age*, **37**, 477, 1956.

BUNN, C.W., *Transactions of the Faraday Society*, **35**, 482, 1939.

CAROTHERS, W.H., *Journal of the American Chemical Society*, **46**, 2226–36, 1924.

—— *Chemical Reviews*, **8**, 353–426, 1931.

—— *Transactions of the Faraday Society*, **32**, 39–49, 1936.

CAROTHERS, W.H. & ARVIN, G.A., *Journal of the American Chemical Society*, **51**, 2560–70, 1929.

FAWCETT, E.W., *Transactions of the Faraday Society*, **32**, 119, 1936.

FAWCETT, E.W. & GIBSON, R.O., *Journal of the Chemical Society*, 384 and 396, 1934.

GOURLAY, J.S. & JONES, M., *British Plastics*, **29**, 446, 1956.

HILL, J.W. & CAROTHERS, W.H., *Journal of the American Chemical Society*, **54**, 1579–87, 1932.

HINDS, L., *Rubber Journal and International Plastics*, 6 February–12 March 1960.

HUNTER, E., *Advancement of Science*, **18**, 171, 1961.

—— *Chemistry and Industry*, 2106–12, 1961.

IZARD, E.F., *Chemical and Engineering News*, **32**, 3724–28, 1954.

NATTA, G., *Modern Plastics*, **34**, 169, 1956.

PERRIN, M.W., *Research*, **6**, 111, 1953.

SWALLOW, J.C., *Endeavour*, **3**, 26, 1944.

—— in *Council News*, Plastics Division Council, I.C.I. Plastics Division, April 1958.

—— *Chemistry and Industry*, 1367, 1959.

WHINFIELD, J.R., *Chemistry and Industry*, **21**, 354, 1943.

—— *Nature*, **158**, 930, 1946.

—— *Endeavour*, **11**, 29–32, 1952.

ZIEGLER, K., *Angewandte Chemie*, **64**, 323, 1952.

—— *Research in the Max Planck Society*, Shell Chemical Co., 1962.

Oxygen Steelmaking

BIBLIOGRAPHIES

Oxygen in Steelmaking 1946–1959, Bibliographical Series no. 22, Iron and Steel Institute, London.

The subsequent period to the present day is covered by an excellent card index held by the Iron and Steel Institute.

BOOKS

ANON., *Basic Open Hearth Steelmaking*, Physical Chemistry of Steelmaking Committee, Iron and Steel Division, A.I.M.E., New York, 2nd edn., 1951.

BASHFORTH, G.R., *The Manufacture of Iron and Steel*, Chapman and Hall, London, 3rd edn., 1964.

BURN, DUNCAN, *The Steel Industry, 1939–59*, Cambridge University Press, 1961.
—— *The Economic History of Steelmaking 1867–1939*, Cambridge University Press, 1961.
CHARLES, J.A., CHATER, W.J.B. & HARRISON, J.L., *Oxygen in Iron and Steelmaking*, Butterworths, London, 1956.
CHATER, W.J.B. & HARRISON, J.L. (eds.), *Recent Advances with Oxygen in Iron and Steelmaking*, Butterworths, London, 1964.
JACKSON, A., *Steelmaking for Steelmakers*, The United Steel Companies Limited, Sheffield, 1960.
—— *Oxygen Steelmaking for Steelmakers*, Newnes, London, 1964.
KEELING, B.S. & WRIGHT, A.E.G., *The Development of the Modern British Steel Industry*, Longmans, London, 1964.
MURRAY, D., *Steel Curtain*, Pall Mall Press, London, 1959.

JOURNALS
CAHN, R.W., *Discovery*, **26**, 41, 1965.
SITTIG, M., *Chemical and Engineering News*, **39**, 92, 1961.
Almost all journals dealing with metallurgy carry papers and articles on various aspects of oxygen steelmaking. Industrial information on the situation in Britain is fairly adequately covered by:
Steel Review, published quarterly by the British Iron and Steel Federation.
Steel Times, formerly *Steel and Coal*, formerly *The Iron and Coal Trades Review*, published weekly by Fuel and Metallurgical Journals Limited, London.

REPORTS
The Iron and Steel Board Annual Reports, H.M.S.O., London.
Development of the Iron and Steel Industry, Special Report, *1961*, The Iron and Steel Board, H.M.S.O., London.
Development of the Iron and Steel Industry, Special Report, *1964*, The Iron and Steel Board, H.M.S.O., London.
Research in the Iron and Steel Industry, Special Report, *1963*, The Iron and Steel Board, H.M.S.O., London.
British Iron and Steel Research Association Annual Reports.
Iron and Steel Annual Statistics, 1963, Iron and Steel Board and British Iron and Steel Federation.

PAMPHLETS AND HANDBOOKS
The Iron and Steel Board—What it is and what it does, The Iron and Steel Board, London, 1964.
The Iron and Steel Institute Handbook 1963, Iron and Steel Institute, London.
The British Iron and Steel Federation, B.I.S.F., London, 1963.
The British Steel Industry, B.I.S.F., London, 1964.
Steel, the Facts, British Iron and Steel Federation, London.

The British Iron and Steel Federation has issued a large number of pamphlets dealing with various aspects of the industry. These include such topics as research, training, recruitment, relations with European countries, technical developments, etc.

SUBJECT INDEX

NAME INDEX